# EXTREME LEADERSHIP

*With Four Case Studies*

by

## Charles Patton

First published in 2022
Revised Edition 2025

Published by Short Mystery Press
Orlando, FL

EXTREME LEADERSHIP

Copyright © 2022, 2024, 2025 by Applied Market Solutions, LLC

Cover Design by Book Design Company

Table of Contents

# PREFACE

This book is more a homage to the work of others than a presentation of startling discoveries by me. In cases where the work of others has been cited, quoted, or incorporated, I have given appropriate attribution. Some historical content comes from antique books published in the 1800s and no longer in print. More recent findings were integrated with proper deference to the originating experts. My contribution lies in identifying the common threads within the experiences of extreme leaders and extracting the common skills they utilize in extreme circumstances.

I want to extend my gratitude to the great historians who diligently captured the lives and tribulations of notable leaders, providing us with invaluable material to consider (see References). I have relied heavily on their original words, as they reflect the language and context of the times far better than I could. In the case studies, their words and ideas are presented with only occasional editing for readability or added observations. In the body of the book, aside from where citations are made to credit other authors, the content is my original work.

I would also like to thank Dr. David Lady, who inspired me to compile the information collected over the years to create this book, and Geoffrey Wilson, who assisted with the first edition by editing and improving the book's readability. Thanks also to Eldo Barkhuizen of Reedsy.com for editing this updated version.

# INTRODUCTION

Extreme situations call for unique measures and exceptional leadership skills. This book delves into those skills, particularly those utilized in extreme circumstances. It aims to guide aspiring leaders and those already in leadership positions in enhancing their leadership abilities, especially in crisis management.

The book features a collection of case studies showcasing extreme leaders and their experiences. While reading all four case studies first is optional, doing so will help you recognize the patterns discussed throughout the book. The goal is to inspire you to search within yourself for your purpose and pursue a Big Vision, as described below.

Effective leadership requires a clear sense of purpose. Joan of Arc, an extreme leader, epitomized this. Known in French as Jehanne d'Arc and preferring the name Jehanne la Pucelle, she famously distinguished between a mere wish and a resolute purpose when asked about her mission to drive the English from French soil and crown Charles V. Leaders possess vision and understand their purpose.

Whether in crises or ordinary times, you can direct your leadership toward good or evil. Both supporters and opponents must scrutinize the morality of a leader's ethics. Studying leadership, even with evil purposes, reveals techniques that worked and those that did not. Common characteristics emerge among leaders in extreme circumstances, regardless of their motives.

Consider Mayor Giuliani's actions immediately after the September 11, 2001, attacks. Hailed for his effective leadership during those horrendous circumstances, Giuliani exemplified extreme leadership with a clear, positive purpose. As you read or listen to this book, consider what made him effective during those critical days and how he has changed in recent years.

On the other hand, Cortes led his followers to invade Mexico, ostensibly to claim it for his monarch and convert its natives but primarily for the nefarious purpose of amassing gold for himself and his sponsors. This book explores the skills used by leaders like Giuliani and Cortes.

Extreme leadership often involves male leaders, as historical opportunities for women in such roles were limited. However, women like Joan of Arc, Mother Teresa, Golda Meir, Margaret Thatcher, and Hillary Clinton have demonstrated exceptional leadership, often with noble purposes. Their contributions are equally significant.

Consider this allegorical tale: A knight errant faces an evil wizard, enduring relentless magic attacks, grave injuries, abandonment, and deprivation. Despite unimaginable hardships, the knight perseveres, exemplifying extreme leadership—committed, focused, unrelenting, and indomitable. These traits define an extreme leader.

Anyone can learn leadership, but extreme situations demand a particular sensitivity to followers' needs, unique courage, and a willingness to accept the consequences. For instance, Eisenhower's D-Day invasion plan carried immense risks. Failure could have destroyed his career and jeopardized the world. He prepared a resignation speech in case of failure, demonstrating the courage to proceed despite uncertainty. This kind of courage is integral to extreme leadership, balancing between the extremes of "doing what must be done" and "damn the torpedoes" – risking everything for the highest stakes against the riskiest odds.

# Chapter 1

# WHAT IS A LEADER UNDER NORMAL CIRCUMSTANCES?

Rather than starting with definitions for leader, leadership, and extreme leadership, let us consider the various characteristics that define leaders. Their traits will naturally illustrate the meanings of these words.

What distinguishes an effective leader from someone simply in a leadership position? It is not just a matter of titles. A person is not a leader until they have led, until they have faced and overcome challenges. Someone might have 'leadership potential,' but they need to be tested to claim the title of leader truly. An ordinary leader will exhibit characteristics such as the following:

### A Leader Steps Forward Willingly When Opportunity Arises or Simply Takes Charge

Becoming a leader starts with being placed in a leadership role or by creating that role oneself. Being placed in a leadership role is often an honor bestowed by someone in power upon someone who has demonstrated qualities such as presence, personality, loyalty, a positive mental attitude, or simply a willingness to volunteer. Other times, a person steps forward and announces their intention to lead.

To truly become a leader, a person in a leadership position must make a difference—not just attempt to make a difference but achieve it. Holding a leadership position alone does not make someone a leader; a leader earns the title through accomplishments while in that role. Most leaders learn to become true leaders through experience, often via trial and error (or trial and success).

Leaders who try but fail repeatedly may lose their following and leadership positions. However, those who learn from their failures gain valuable lessons. They may find new leadership opportunities, or they may choose to stay and fight through the issues. The key is to never let failure be the end of the road.

Often, even those removed from a leadership position due to failure can rise again in another leadership role and become excellent leaders because of the lessons learned from those failures. Abraham Lincoln is a prime example; see the following table. Fehrenbacher, 1992)

| YEAR | FAILURES or SETBACKS | SUCCESSES |
|------|---------------------|-----------|
| 1832 | Lost job<br>Defeated for state legislature | Elected company captain of Illinois militia in Black Hawk War |
| 1833 | Failed in business | Appointed postmaster of New Salem, Illinois. Appointed deputy surveyor of Sangamon County |
| 1834 | | Elected to Illinois state legislature |
| 1835 | Sweetheart died | |
| 1836 | Had nervous breakdown | Reelected to Illinois state legislature (running first in his district)<br>Received license to practice law in Illinois state courts |
| 1837 | | Led Whig delegation in moving Illinois state capital from Vandalia to Springfield<br>Became law partner of John T. Stuart |

| | | |
|------|------------------------------------------|--------------------------------------------------------------------------------------------------------------------------------------------------|
| 1838 | Defeated for Speaker | Nominated for Illinois House Speaker by Whig caucus. Served as Whig floor leader.<br><br>Reelected to Illinois House (running first in his district) |
| 1839 | | Chosen presidential elector by first Whig convention |
| 1840 | | Argues first case before Illinois Supreme Court Reelected to Illinois state legislature |
| 1841 | | Established new law practice with Stephen T. Logan |
| 1842 | | Admitted to practice law in U.S. District Court |
| 1843 | Defeated for nomination for Congress | |
| 1844 | | Established own law practice with William H. Herndon as junior partner |
| 1846 | | Elected to Congress |
| 1848 | Lost renomination | Chose not to run for Congress, abiding by rule of rotation among Whigs |
| 1849 | Rejected for land officer | Admitted to practice law in U.S. Supreme Court. Declined appointment as secretary and then as governor of Oregon |

| 1854 | Defeated for U.S. Senate | Elected to Illinois state legislature (but declined seat to run for U.S. Senate) |
|------|--------------------------|----------------------------------------------------------------------------------|
| 1856 | Defeated for nomination for Vice President | |
| 1858 | Again defeated for U.S. Senate | |
| 1860 | | Elected President |

A *leader leads* even when not in a formal leadership role—being willing to take charge, accept responsibility, and be capable of managing any situation responsibly whenever the opportunity arises. Most leaders have an internal drive to lead; they want to take charge out front. Others prefer to follow, which is perfectly fine because only one can lead, and not everyone is cut out for the stresses of leadership. Some would just as soon stay back and observe what others do. As the adage goes, "Lead, follow, or get out of the way."

Audie Murphy, a fifth-grade dropout from a poor family in a small Texas town, faced rejection from the U.S. Army, Navy, and Marines due to his young age, short stature, and underweight condition. Finally accepted into the Army as a private, he had to fight for the chance to see combat. Once in battle, he emerged as a World War II hero—the most decorated soldier, earning the Medal of Honor. After the war, he successfully transitioned to become a movie star despite starting out as a follower.

The only person who will never lead is the one who stays home and avoids any opportunity to become more.

### A Leader Has Wisdom

Willingness to lead is often mistaken for leadership wisdom. An eager individual who aspires to lead will readily volunteer whenever the opportunity arises, initiating the process of gaining experience. Consequently, a 'new' leader may exhibit energy and strong organizational skills, which can sometimes be misinterpreted as effective leadership.

What is truly essential, however, is leadership wisdom. A seasoned leader possesses deep insights into the environment, people, and subject matter expertise crucial for success. This wisdom typically develops from extensive experience gained in challenging situations. Yet, leadership wisdom only manifests with initial willingness to lead. Combining wisdom and willingness is vital but not sufficient for successful leadership—it also requires motivating followers.

### A Leader Motivates

A leader motivates others to achieve an overarching vision, often defined by a series of goals known as their Big Vision, representing the ultimate purpose. Motivation stems from these articulated goals, especially when leaders gain agreement from their followers. Creating positive motivation is a skill anyone can learn, but not everyone can master it. Based on my experience, the following factors significantly influence motivation and foster commitment:

EXTREME LEADERSHIP

1. **Belief in the goals**: A team must believe in the worth of its goals. Teams will willingly make sacrifices against overwhelming odds for causes they believe in, no matter how difficult or risky. A leader must exert significant effort to communicate, ensure understanding of, and secure acceptance of the goals among the followers.

2. **Energy**: Motivation requires energetic leadership. The leader's enthusiasm fuels motivation. A lethargic environment fails to inspire or motivate.

3. **Commitment**: Consistent leader commitment to the mission is crucial. Followers notice and mirror the leader's level of commitment. For instance, leaders should avoid asking people to work late every day while leaving early themselves.

4. **Inclusion**: Team members must feel like insiders, receiving all information and feedback—both positive and negative. Trust, loyalty, and fair treatment sustain inclusion and commitment. Communication must be comprehensive and timely.

5. **Recognition**: Individuals crave recognition, both individually and as team members. Effective leaders celebrate successes and unite everyone during impending failures. Leaders should celebrate achievements at every opportunity, fostering a culture of extraordinary effort. This recognition acknowledges individual and team contributions and makes everyone feel valued and appreciated. However, leaders should be mindful not to single out individuals for recognition when others have made equal sacrifices without acknowledgment.

6. **Rewards**: Teams need to believe there will be a pay-off for success, whether material (like promotions) or non-material (such as pride in achieving a noble cause). Celebrating success at the mission's end is vital for overall follower satisfaction. The promise of rewards motivates the team and gives them hope for a brighter future, reinforcing their commitment to the mission.

Experience and the history of extreme leaders have demonstrated that job security, good pay, and comfortable working conditions are not essential for achieving high motivation. Some of the greatest accomplishments have been realized under the most adverse conditions. For instance, Neil Armstrong, Mike Collins, and Buzz Aldrin's mission to the moon and Columbus's accidental discovery of the Americas instead of a new route to the Far East highlight how people are willing to sacrifice for a cause they believe in.

The strongest motivation arises when extreme leaders empower their followers with authority and responsibility to develop initiatives and plans that align with the Big Vision. While the extreme leader and senior leaders may establish the Big Vision and outline key goals, handing the goals over to the followers achieves maximum buy-in. Leaders provide guidance and resources as needed, intervening only to maintain focus if necessary. Leaders encourage ownership and commitment by delegating the task of refining goals and devising initiatives, schedules, and deadlines to their followers. Followers who contribute to the planning process are more likely to dedicate themselves fully to executing the plan.

### A Leader Is Willing to Take Risks

A leader possesses courage and is willing to take risks to drive change. Those in leadership who avoid risks will eventually be seen as ineffective. Leaders who take calculated risks and succeed are often celebrated, while those who take risks and fail may still be acknowledged for their bold attempts.

Furthermore, leaders who spend excessive time assessing risks often falter, just as those who act hastily without adequate assessment. Successful leaders operate within a spectrum, finding a reassuring balance between having sufficient information and acting decisively despite uncertainties.

General George S. Patton leaned towards one end of the spectrum, often making swift decisions with limited information, although he faced constraints such as the pace of his tank's fuel supplies. In contrast, General Bernard Montgomery, according to General Patton and sometimes General Eisenhower, took a more cautious approach, meticulously assessing risks and planning missions at what Patton perceived as a slower pace. Despite their differing styles, Patton and Montgomery emerged as victorious generals during the war. They achieved this success by operating within the bounds of reasonable risk-taking.

### A Leader Must Know How to Innovate

A leader must be innovative, especially when the team

encounters challenges or faces imminent deadlines. In these critical moments, a leader's experience and wisdom are crucial for finding solutions and ensuring success.

Moreover, a leader understands how to foster creativity within the team and from external sources. Skillfully conducting brainstorming sessions often yields valuable insights and improvements, even when conducted spontaneously under pressure. The leader knows when to initiate and conclude a brainstorming session, how to guide the discussion effectively, and how to refocus the team afterward to integrate new ideas into the existing strategy. However, the leader always bears responsibility for the decisions made, regardless of the idea's origin.

## A Leader Sets the Example and Sets the Bar

A leader not only sets high expectations and goals for others but also actively contributes to the effort. A good leader does not merely oversee the execution of the plan; their efforts set the standard that others are expected to meet or exceed.

---

When Joan of Arc led her armies into battle, she rode out in front of them carrying a banner that symbolized their mission—to drive the English from French soil. Similarly, General Patton could often be found at the forefront, personally inspecting his troops and field commanders. Being with the soldiers at the front is what followers expect of a great leader.

---

Effective leaders do not delegate all difficult tasks but instead retain some challenging responsibilities for themselves to demonstrate they are not exempt from what they expect of others. Good leaders are willing to perform any task they delegate to their team.

A strong leader actively contributes to the team's efforts while maintaining a balanced perspective on the overall endeavor. They step in when necessary, working alongside their team to push forward. A good leader strives to avoid signaling superiority over any team member; teamwork thrives on equality. Egos have no place in a team-driven effort.

## A Leader Provides Rewards

In addition to the effect of rewards on motivation described earlier, a leader often gains initial commitment through promises of "loot"—something valuable to be gained at the end. Cortes, for instance, offered his men a share of the gold he expected to find. This "loot" is not always material; it can be the satisfaction of doing something good for others. General Patton promised his men victory by "making the other dumb bastard die for his country." Rewards can also be the mental satisfaction of accomplishing the goals set by the team. A sense of accomplishment can be profoundly satisfying.

When I was young, I spent my summer breaks working at the Smith family dairy farm outside the small town where I grew up in Illinois. Reflecting on those times later, I realized the work was incredibly fulfilling. The labor was physically demanding and exhausting, with days starting at 3:30 AM and ending at 10:30 PM, six days a week. Financially, it was minimally rewarding, paying only $40 per week (about $0.50 per hour, equivalent to $5.00 per hour in 2024). Despite these challenges, I recalled the experience as fulfilling because, at the end of each day, I could see the tangible results of my efforts—the stack of hay bales, the silo full of silage, the hundreds of gallons of milk harvested, the cows fed, and so on. This immediate feedback was extraordinarily satisfying. It motivated me to rise to the challenge day after day and never tire of the work.

A good leader understands that seeing the results of one's efforts is highly motivating. Therefore, they will highlight small successes as often as the large ones, expressing appreciation and fostering a sense of fulfillment and pride in a well-done job.

### A Leader Maintains a Positive Attitude and Is Visible

Leaders maintain a can-do attitude and stay visible to their followers, superiors, and associated third parties. Norman Vincent Peale's 1952 book The Power of Positive Thinking popularized this concept, which continues to influence people today (Peale, 1996).

> "Formulate and imprint a mental picture of yourself succeeding indelibly in your mind. Hold this image tenaciously and never allow it to fade. Your mind will work to develop this picture. Do not create obstacles in your imagination" (Peale, 1996).

Positive-minded individuals are effective, whether they are leaders or followers. Success in either role often stems from a foundation of positive thinking.

### Other Leaders' Positive Characteristics

In his book The Art of the Leader, Dr. William A. Cohen identified several personal strengths of leaders. These strengths, which also apply to extreme leaders, include the following (Cohen, 1990, pp. 79–86):

1. **Strength of Mind:** Demonstrates self-control, self-confidence, courage under stress, and healthy instincts derived from hands-on experience.

2. **Firmness:** Holds stable views rooted in consistent thought, reflection, and proven principles. This firmness is balanced with caring, fairness, dependability, and trustworthiness.

3. **Sound Decision-Making:** Merges thoroughness, quickness, and decisiveness to form a robust decision-making process. A proficient decision-maker in a crisis is an asset, adept at averting unnecessary risks and errors with swift and effective responses.

4. **Knowledge:** Possesses knowledge of opponents, current issues, and the virtues, strengths, and weaknesses of those under their command.

5. **Creativity:** Shows the ability to devise new solutions to old problems or innovative solutions to new challenges.

6. **Endurance:** Demonstrates the ability to endure prolonged efforts bolstered by meticulous preparation. This resilience is critical in professional settings to ensure tasks are completed thoroughly and diligently.

7. **Staunchness:** Resilient in recovering from severe setbacks, bolstered by mental and physical fitness.

In Dr. Cohen's list above, many traits are driven by the wisdom of experience. Items one through four and seven depend heavily on experience. Item five might also benefit from understanding what has worked or failed in the past.

Dr. Cohen also wrote that leaders need to be willing to take charge, accept risk, and responsibly manage the power delegated to them. Leaders must know how to set high expectations and goals for others.

To his list, I would add that leaders with honorable purposes need honesty, integrity, trustworthiness, ethical behavior, and honor. These additional traits are important in life and essential for leaders under both normal and extreme circumstances. Leaders often have to make unpopular decisions, such as separating ineffective or dissenting followers or going against their followers' passions. However, based on their wisdom and experience, these decisions are often necessary for the greater good.

In these cases, if leaders have the respect of their followers and are known to be honest, ethical, earnest, honorable, and transparently sincere, then their followers will accept tough decisions. The role of followers in supporting the leader's tough choices is crucial. Usually, when tough decisions are made, the followers already know in their hearts that they are necessary. Without the respect of followers, making hard decisions can break down morale and destroy motivation.

## Leaders Learn to "Win Friends and Influence People"

In his groundbreaking book How to Win Friends and Influence People, Dale Carnegie wrote that a leader's job often involves changing people's attitudes to influence their behavior. Some of his suggestions for accomplishing behavior changes include the following (Carnegie, 1936):

1. Begin with praise and sincere appreciation.
2. Address people's mistakes indirectly.
3. Discuss your own mistakes before criticizing others.
4. When giving instructions, it is often more effective to ask questions instead of issuing direct orders. This encourages participation and engagement, which can lead to better communication.
5. Allow the other person to save face.
6. Praise even the slightest improvement and acknowledge every progress.
7. Give the other person a fine reputation to aspire to.
8. Use encouragement and make faults seem easy to correct.
9. Make the other person feel happy about doing what you suggest.

The characteristics mentioned above form the foundation of effective leadership. These are learned skills—you do not have to be born with them. Understanding these skills is not enough; anyone aspiring to be a good leader should actively practice them while seeking opportunities to step forward and lead.

If good leaders possess these traits, what characteristics do non-leaders exhibit?

# Chapter 2

# NON-LEADER BEHAVIORS

Non-leaders come in various types. Some prefer to be followers because they believe they lack leadership ability. Others may know from self-evaluation or experience that they do not possess leadership skills. Fear of potential failure or the desire to avoid the stress that comes with leading also deters some individuals. There are many reasons for not wanting to be a leader. However, not being a leader and unwillingness to be a dedicated follower can create problems for the team.

Too often, those who resist being led or refuse leadership become critics of those who lead. Some may be indifferent or complacent about the level of commitment around them, while others might be malicious or attempt to undermine efforts. Often, they are jealous of the privileges that come with leadership without being willing to make the corresponding sacrifices. When such behavior is identified, it is important to take action. Those who are indifferent, overly fearful, jealous, or disruptive should be removed from the organization as soon as they are identified. This proactive approach empowers the team to maintain a healthy and productive environment.

### Non-leaders/Non-followers Avoid Risks

A non-leader/non-follower demands job security and guaranteed pay, avoiding risk regardless of the endeavor's success or failure. They are strong advocates of seniority and oppose pay-for-performance unless it is above their guaranteed pay and requires minimal effort.

### Non-leaders/Non-followers Seek Definition

A non-leader/non-follower needs clear rules, defined expectations, and stable job responsibilities. They overemphasize process, chain of command, job descriptions, assigned duties, and policies. Such individuals are uncomfortable with ambiguity and rapid change and resist it at every opportunity.

It is crucial to distinguish between two types of individuals: those who are neither leaders nor followers and seek precise definitions of responsibilities to limit their workload and subordinate leaders who seek clarity to understand the boundaries of their leadership roles. While the latter is essential for optimizing the assignment of responsibility, the former aims to avoid responsibilities altogether.

## Non-leaders/Non-followers Complain Often

Non-leaders/non-followers tend to be critical of others, especially the overall endeavor. They rarely offer solutions to the problems they complain about, yet this does not deter them from constant criticism. Their negativity and lack of constructive input can drain the positivity and enthusiasm of those around them.

## Non-leaders/Non-Followers Produce Routine and Repetitive Output

Non-leaders/non-followers tend to be plodders, engaging in repetitive tasks requiring minimal thinking and offering slight variation. They resist taking on additional work or new tasks, often claiming that their current workload does not permit it, even when it is not particularly demanding. Individuals who consistently arrive and leave on time, showing little passion for the organization's priority initiatives, are typically non-leaders/non-followers.

## Non-Leaders/Non-followers Lack Clear, Steady Focus

Non-leaders/non-followers can be easily distracted by marginal work assignments or colleagues who prefer to chat and gossip. They may require daily or frequent supervision to stay focused and produce consistently, or, if sufficiently entrenched, they may deliver the same mediocre performance day after day. When asked to assist in other areas, they often find reasons why they cannot help or complain that it is someone else's responsibility.

In some cases, especially in large organizations, non-leaders/non-followers may be viewed as the grease that lubricates the organization's wheels. While they may be needed for large organizations to run smoothly, they are not the essential parts of the organization and would struggle to survive in smaller entrepreneurial environments. They must not be allowed to affect the morale and performance of key followers who are crucial to the mission.

Leaders must recognize these characteristics among followers, as failure to do so can lead to a leader's downfall. Even good leaders may fail if they exhibit some of these traits.

> *Warning:* Avoid placing individuals with these characteristics and behaviors into leadership or other critical positions.

Having non-leaders/non-followers in the organization is just one of the many factors that can contribute to a leader's failure.

Chapter 3

# WHY LEADERS FAIL

In his insightful work, Dr. Cohen also identified several reasons why leaders fail (Cohen, 1990, p. 246). Understanding these reasons is crucial for anyone aspiring to lead effectively. Leadership is a challenging role that requires skills, experience, and the right mindset. Even the most capable leaders can stumble if they fall into certain pitfalls. By recognizing and addressing these common issues, leaders can better navigate their responsibilities and avoid failure. Here are the key reasons Dr. Cohen outlined for why leaders may fail.

### Cannot Get Along

When leaders cannot get along on one or more levels—whether with superiors, subordinates, or peers—they are likely to fail. This inability to get along can sever access to critical inputs, resources, and essential information. Furthermore, it can cause others to oppose the leader or the primary mission, undermining overall success.

### Does Not Adapt to Change

In my experience, this is a major reason for leader failure. Being rigid, resisting change, sticking to the original plan despite shifts in its foundation, and failing to respond to critical changes in the surrounding environment can lead to a leader's downfall.

> Dr. An Wang, an engineer and inventor, founded Wang Laboratories in 1951, revolutionizing the computer industry. His company once had a dominant position in the word processing market with its WPS system. However, he failed to adapt to the emergence of the PC, Windows, and Microsoft Word, ultimately losing control of the market to Microsoft. Similarly, IBM lost the PC hardware battle to Dell because it could not, or would not, match Dell's costs. Additionally, IBM's PC operating systems lagged what Microsoft offered. Both companies were once poised to dominate their respective markets but failed to adapt, leading to their decline.

Failing to adapt to change is the polar opposite of what extreme leaders do. Extreme leaders not only respond rapidly to change but also drive it proactively.

**Times Decisions Wrongly**

In my experience, timely decision-making is critical to effective leadership. Being slow to respond when a decision is needed is poor leadership. Conversely, deciding too hastily can also lead to a leader's downfall. Sometimes, taking time to decide is correct, while at other times, a quick decision based on instincts is necessary. Judging the appropriate timing for decisions is a crucial skill that comes with considerable experience.

Dr. Cohen identifies fear of action, an inability to resolve issues, and a reluctance to assess or accept risks as significant weaknesses that contribute to leadership failure.

**Is Preoccupied with Himself**

When ego takes precedence over purpose, failure is inevitable. Leaders who focus solely on WIIFM (What is in it for me?) will lose followers as self-interest becomes apparent or sensed.

**Unable to Rebound – Inability to Weather a Setback**

Too often, leaders become overly invested in their ideas, enamored by them, and overconfident. When this happens, they will miss signals indicating the need for change or impending disaster, much like the Cannonball Express. This speeding passenger train crashed into a stopped freight train in Dunlop, Virginia, in June 1903.

Despite overwhelming evidence that a course change is necessary, egocentric leaders may insist they know what is best and persist on the old path despite apparent dangers. Such leaders are poor listeners, ignoring others who try to warn them about the issues.

### Fails to Allow Positive Solutions Failure to Overcome the Failure Reasons

There are natural balancing tendencies in organizations. When a problem or crisis arises, other members of the organization, external sources, and superiors can often assist if a leader is open to asking for and accepting help. If not too frequently. Recognizing when help is needed and seeking it should never be seen as a weakness in leadership but rather as a crucial strength.

However, some higher-level leaders maintain an attitude that they do not want to be bothered with problems unless they devise a solution. This approach reflects weak leadership. A senior leader, as a key figure in the organization, should be ready to assist direct reports when they need help, instead of prematurely assuming they are failing and need replacement. This supportive approach not only strengthens the team but also fosters a culture of open communication and trust.

> Many entrepreneurial bosses mistakenly believe that dismissing and replacing employees is the best solution to challenges, echoing my friend Larry Darrah's observation that "the smart guys all work somewhere else."

Moreover, sometimes, a situation resolves itself if you exercise patience and are willing to let the problem unfold. Leaders who close themselves off to these possibilities increase the likelihood of failure and possibly ensure it.

### Fails to Set, Monitor, or Enforce Deadlines

Individuals without timelines lack focus—the two go hand in hand. Beyond setting deadlines, consistently monitoring, and enforcing them are essential leadership skills. Leaders who neglect these practices are destined to fail. The inability to control project scope creep or allow deadlines to slip without thorough investigation and reassessment of the project timeline leads to unforeseen challenges.

### Fails to Correct Errors as Soon as They Occur

Responding slowly or not at all when mistakes occur is sure to doom a leader. When a leader makes a mistake, the sooner it is acknowledged and corrected, the better. If a subordinate makes a mistake, an investigation is warranted to ensure proper processes are in place to prevent it from happening again.

> My directions to my direct reports have always been: "You can make a mistake, and we will address it but do not make the same mistake twice and avoid making too many different mistakes."

### Fails to Accept Responsibility

The worst mistake leaders can make is avoiding taking responsibility for their actions' outcomes and consequences. If an endeavor fails, the last thing a leader should do is claim to be a victim, blame others, blame their followers, or blame circumstances. A leader can almost always do something differently to prevent failure. With proper planning, a leader may even mitigate Acts of God.

Consider the catastrophic consequences of the tsunami that struck Japan's nuclear power plants in 2011. While the tsunami could not have been prevented, the disaster's impact could have been mitigated. The protective barriers around the plants could have been built higher and stronger, and additional engineering measures could have ensured better backup power options and other safeguards.

As the adage goes, "Hindsight is 20/20," but there is always room for improvement in every case where disaster strikes, such as the New Orleans levee breaches. Preparation could have been better, engineering more thorough, money could have been spent, and the response could have been quicker. Each of these "could haves" falls under the responsibility of leaders who either failed to do their job or were prevented from doing so.

### Fails to Muster What Is Needed to Succeed

Success requires experience, resources, energy, enthusiasm, dedication, concentration, and execution. A leader's failure to mobilize these elements effectively will lead to failure. In some instances, inaction or incorrect actions can result in loss of life. Decisions that involve life-and-death stakes are particularly challenging to assess and make.

Imagine President Barack Obama sitting in the Situation Room, deciding to send Seal Team Six into Pakistan to kill Osama Bin Laden. Although our Special Forces teams are among the best in the world, they are not invincible, and the loss of any member is a profound blow to their leaders. We have experienced the tragic loss of helicopters full of soldiers in the Middle East far too often. The decision carried immense risks, not only politically but also in terms of foreign policy, yet Obama made it.

Courage and access to the best information are essential for making critical and dangerous decisions. If the current leader is failing, there may be times when a new leader must emerge to step into the role, whether by the request of superiors or followers or by their initiative. Failure creates opportunities for an emerging leader to take over the reins. However, it is important not to step up prematurely to depose a leader without the support of those who decide on leadership. Otherwise, if the original leader regains the support of those who initially placed them in power through persuasion or new successes, they may seek revenge, figuratively or literally.

To be successful, leaders need followers, and exceptional

leaders need highly committed followers. These followers are not just a support system, but a significant factor in the success of a leader. Their commitment and dedication can empower a leader to achieve great things.

Chapter 4

# WHO ARE THE FOLLOWERS?

If you consider the accomplishments of historic extreme leaders, not one of them could accomplish what they did alone. Edmund Hillary climbed Mount Everest first but Tenzing Norgay, his Sherpa, was at his side. When Neil Armstrong stepped on the moon, he could not have done so without the help of Michael Collins piloting the Command Module, Buzz Aldrin landing the Lunar Module on the moon's surface, and thousands of people at NASA who built their spaceships and guided them to the moon and back. And, if you consider Christopher Columbus, Hernando Cortes, Mahatma Gandhi, and other extreme leaders throughout the years, you might think that they accomplished what they did by stirring into a lather a large batch of bumbling, unthinking followers. But that thought would be wrong. Followers are not mindless – they often have very valuable skills and a high level of competence in applying those skills. They usually require little direction and are usually very loyal and as committed to accomplishing the vision as the leader. They may or may not have the same vision as the leader, nor the same ability to influence others, but can be just as committed to the mission and just as intrepid as the leader, fighting side by side in the midst of the fray. And sometimes they may even sacrifice their lives to protect the leader.

## Why Do Followers Follow?

Whether a participant in the quest is a follower or an intermediate leader, people act based on their reasons, which may not align with the extreme leader's motivations. Followers might volunteer in hopes of gaining fame or fortune, advancing their careers by taking on more responsibility, or out of friendship and love for the leader. While followers' reasons differ from the leader's higher purpose, spiritual inspiration, or noble goals, some followers may share equally noble or completely aligned motivations.

However, regardless of their motives, followers must align their interests with the leader's or at least identify with the leader's interests. By doing so, they effectively align their future with the leader's.

## How Are Followers Attracted?

Followers may be drawn to a leader who steps forward to address an issue they care about. They might be attracted just because the leader asked for help or because they believe in the vision and sympathize with the leader's goals. Initial followers often attract more followers by spreading the word. For instance, Martin Luther King's peaceful demonstrations inspired many to join his cause.

Followers are most likely to be drawn to a leader who recognizes and begins to address a severe problem or opportunity. Public examples of the leader's commitment, such as helping those in need, further galvanize support and attract additional followers.

## How Should Followers Be Managed?

Followers should be managed like any effective work team: receive clear assignments, be well-informed, and be recognized for their accomplishments.

> Robert Townsend, author of *Up the Organization* and CEO of Avis Car Rental in the 1960s, stated, "True leadership must benefit the followers, not enrich the leaders."

Followers can be organized into teams and positively energized through competition with other teams. Individual followers should be praised in public and disciplined in private.

During extreme circumstances, harsh and even cruel behavior by the leader may become necessary to maintain discipline and enforce essential policies (e.g., Cortes ordered that women were not to be raped). Such actions will be judged later as necessary and acceptable if the endeavor succeeds. However, if the effort is unsuccessful, these actions might be considered unnecessary and excessively harsh. If a harsh leader is unsuccessful, the behavior may be pointed to as one of the reasons for failure, may have been one of the causes of failure, and could even lead to prosecution for unlawful acts.

## How Are Followers Motivated?

As previously mentioned, rewards should be delivered at the end of the mission and whenever opportunities for celebration arise along the way. Information sharing is another crucial motivator. People want to feel like insiders and be essential to the solution. If they are not kept informed, they will feel expendable. Treating them as unique, such as seeking and respecting their opinions, is highly motivating.

Years ago, I helped lead several "Town Hall Meetings" in a company experiencing morale issues. The purpose of these meetings was to address and resolve the concerns. The ground rules were straightforward: decisions that could be made immediately would be made at the meeting. Decisions requiring further research would be deferred, but a list would be posted with dates indicating when each decision would be made and announced. If a request could not be fulfilled because it did not align with the company's strategies or policies, attendees would be informed immediately. These meetings resolved many pressing issues, and attendees gained a better understanding of why some decisions could not be made on the spot. They also saw that deferred decisions were being actively considered. The meetings and follow-up efforts significantly alleviated the attendees' frustrations and improved morale.

Resolving followers' issues promptly, openly, and honestly is an excellent way to build trust and improve morale.

### How Are Followers Handled When Discipline Is Needed?

Rules must be clear, and violations must be addressed swiftly, fairly, and consistently. Sanctions should be applied when necessary, and encouragement should be provided when needed. Recognition and rewards, as previously discussed, balance the occasional need for toughness.

> Years ago, my managers labeled a discipline technique I used as "The Sandwich" because I always privately presented criticism of an employee's performance sandwiched between positive reinforcement. First, I would remind them of their value and acknowledge their good deeds. Then, I would address the specific issue, providing clear direction on what needed improvement. Finally, I would express my confidence in their ability to correct mistakes and continue performing well.

Discipline is occasionally necessary, but the goal is to ensure that the undesirable behavior is not repeated.

**Other Follower Treatment**

Followers should be treated as if they are generals in their own right. They will act responsibly when they feel responsible and are recognized for their responsibility. Some will rise to the challenge and operate independently, while others may not. Extreme leaders need to understand their followers' skills and capacities, and this knowledge is best gained by testing them with responsibility.

Not all followers will respond as generals. Some will need simple and explicit orders, while others may lack independence. Most followers have an innate willingness to obey orders and a desire to please, and all will respond with praise, rewards, and recognition.

These are the traits of followers. The previously described traits of a leader are found in extreme leaders but to a greater degree and with additional characteristics.

Chapter 5

# CHARACTERISTICS OF AN EXTREME LEADER

When examining the characteristics of extreme leaders, it became clear that they possessed all the traits of an ordinary leader but with exceptional commitment, focus, and drive. They operate on a higher plane than typical leaders, demonstrating unparalleled dedication and motivation.

Extreme leadership is the art and skill of guiding and inspiring a team in high-stakes, unpredictable situations where rapid and decisive actions are essential. It involves building and maintaining trust, fostering open communication, and instilling team confidence, ensuring everyone is motivated and aligned toward common objectives despite the severe risks involved. This form of leadership requires a unique blend of resilience, adaptability, emotional intelligence, and the ability to make life-altering decisions under intense pressure, all while cultivating a strong, committed following.

Extreme leadership differs significantly from conventional leadership in its application, context, and demands. The following additional characteristics distinguish such leaders.

## The Mindset of an Extreme Leader

The mindset of an extreme leader has a distinct set of psychological traits that enable them to excel in high-pressure environments. This mindset's core is resilience—the ability to recover from setbacks, adapt to changing circumstances and lead effectively despite significant stress and challenges. Extreme leaders exhibit a high degree of emotional intelligence, allowing them to manage their own emotions and understand and influence the feelings of others. They maintain a calm and focused demeanor, crucial for making clear-headed decisions amidst chaos.

Additionally, these leaders possess an unwavering commitment to their mission and team and a profound sense of responsibility and ethical clarity that guides their actions even in the most trying conditions. This mindset equips them to handle immediate crises and inspires confidence and steadfastness among their followers. Consequently, they forge a team that is resilient, responsive, and deeply aligned with the leader's vision.

### High Resilience

Extreme leaders demonstrate an exceptional ability to rebound from setbacks and maintain their composure and clarity of thought under extreme stress. Learning resilience is a dynamic process cultivated through personal experiences, self-awareness, and deliberate practice. Embracing challenges as opportunities for growth shifts your mindset toward resilience, allowing you to view obstacles as chances to learn and advance.

Building a strong support network of family, friends, colleagues, and mentors is crucial for providing essential emotional support and practical advice during tough times.

44

Equally important is the proactive practice of stress management techniques such as mindfulness, meditation, or physical activities, which help maintain mental health and emotional equilibrium. Setting realistic goals and breaking larger objectives into manageable tasks can prevent feelings of being overwhelmed and foster a sense of accomplishment.

Reflecting on harrowing experiences allows for learning and preparation for future challenges. Maintaining perspective helps manage stress by focusing on long-term outcomes rather than temporary setbacks. Cultivating a positive outlook encourages the expectation of good things and helps visualize success. Focusing on aspects within your control rather than external circumstances reduces feelings of helplessness. Lastly, engaging in self-discovery to understand your deeper motivations and values can significantly enhance your resilience, better preparing you to face and adapt to life's challenges.

### Adaptability

They can adjust strategies quickly in response to changing environments and unexpected challenges, demonstrating a high degree of flexibility. Learning adaptability involves not just accepting but embracing change as a natural and positive aspect of life. It is about stepping out of your comfort zone to gain new experiences. Enhancing problem-solving skills through diverse challenges and maintaining calm through effective stress management techniques are crucial.

Staying informed about new information and current events prepares you to anticipate and adapt to changes more adeptly. Practicing flexibility in thought and action and building confidence through achievable goals fosters adaptability. Additionally, seeking feedback and learning from mistakes are essential for continual improvement. Cultivating a mindset of lifelong learning ensures ongoing growth and the ability to handle new and unexpected situations easily. By integrating these approaches, you can develop a robust ability to adapt, which is crucial for successfully navigating various life and professional scenarios.

### Decisiveness Under Pressure

Extreme leaders must make fast, often high-stakes decisions with incomplete information. Being decisive under pressure is a vital skill that requires cultivating a clear and focused mindset, even in stressful situations. Developing this capability involves regular exposure to high-pressure scenarios, where you can practice making quick decisions and observe their outcomes. Strengthening your knowledge base and staying informed about your field can provide the confidence to make swift decisions.

Enhancing your problem-solving skills through activities that challenge your critical thinking and require swift action is crucial. Establishing a support system of mentors and peers is key, as their guidance and feedback are invaluable in refining your decision-making process. Training yourself to think ahead and anticipate potential scenarios aids in evaluating options quickly and effectively under stress. Focusing on self-awareness and stress-management techniques, such as deep breathing or meditation, can calm the mind and enhance clarity, making it easier to make critical decisions when they matter most.

## Crisis Communication Skills

They excel in communicating clearly, concisely, and quickly, ensuring that critical information is accurately conveyed and understood in urgent situations. Mastering practical crisis communication skills involves honing the art of clear, concise, and transparent messaging during high-pressure situations. This skill set can be developed through focused training, including role-playing scenarios and crisis-simulation exercises, which provide real-time practice in crafting and delivering messages under stress.

Understanding the importance of timely communication and audience-specific messaging is crucial. This skill means tailoring information to different stakeholders, from the public to internal teams, ensuring it is accessible and actionable. Building a solid foundation in emotional intelligence helps communicators empathize with their audience, manage their emotional responses, and adjust their communication style based on feedback.

Regularly studying past crisis communication cases offers valuable insights into successful and unsuccessful strategies, further honing your ability to navigate and communicate effectively during emergencies. Lastly, always be prepared with a communication strategy that includes predefined protocols for different types of crises, ensuring that the response is swift and organized when a crisis does strike.

## Emotional Intelligence

Managing their emotions effectively and reading and responding to the emotions of others is crucial, especially in high-pressure environments where team morale and cohesion are at risk. Developing emotional intelligence (EI) involves a deep understanding of both your own emotions and those of others. This skill can be cultivated through self-reflection, where regularly assessing and analyzing your emotional responses helps identify patterns and triggers in your behavior. Engaging in listening and empathy exercises enhances your ability to perceive and relate to the emotions of others, which is crucial for effective interpersonal interactions.

Participating in training programs and workshops focused on EI skills, such as conflict resolution and mindful communication, provides practical tools and techniques for managing emotions constructively. Seeking feedback from peers and mentors can offer valuable insights into how your emotional behavior impacts others. Reading extensively about emotional intelligence theories and applications also broadens your understanding and provides strategies for applying EI principles in daily life.

Practicing mindfulness through meditation or yoga is a powerful tool for improving emotional regulation. This, in turn, helps you maintain emotional balance and enhance overall emotional intelligence. Mindfulness is a practical technique that can be incorporated into your daily routine, contributing to your personal and professional development.

### Courage and Moral Integrity

Extreme leaders often face ethical dilemmas and must make tough choices that require courage and a solid moral compass. Developing courage and moral integrity involves a deeply personal process of aligning your actions with ethical principles and values, even in the face of adversity. This growth begins with self-reflection, where individuals assess their values, identify what they stand for, and commit to acting consistently with these beliefs.

Reading biographies of leaders known for their integrity and moral courage can inspire and provide practical examples of these traits. Engaging in discussions and debates on ethical dilemmas in educational settings or informal groups sharpens your ability to navigate complex moral issues. Mentorship plays a crucial role, as mentors can guide, challenge, and support individuals in making tough decisions that reflect courage and integrity.

Setting personal challenges that require stepping out of comfort zones and facing fears can build courage incrementally. However, the practice of transparency and taking responsibility for actions truly empowers an extreme leader. Standing up for what is right, even when it is difficult, reinforces moral integrity in everyday life. Over time, these consistent practices embed courage and integrity as integral components of character, enhancing the ability to act boldly and ethically under pressure.

# How Extreme Leaders Differ from Conventional Leaders

### They Forge Their Skills in Crisis

Extreme leadership is often exercised in crises such as natural disasters, military conflicts, or high-stakes business turnarounds. These environments are less structured and more chaotic than conventional leadership settings.

### They Make Risky Decisions

The consequences of extreme leadership decisions are immediate and significant, often involving serious risks to life and property. In contrast, conventional leadership decisions typically involve less immediate risk and allow more time for reflection and consultation.

### They Tolerate and Manage Stress in Others

While all leadership roles involve stress, extreme leadership demands managing acute stress and maintaining high performance in situations where failure can have dire consequences.

### They Rely on Intuition

Due to the need for rapid decision-making, extreme leaders rely more heavily on intuition than data-driven approaches, which are more common in conventional leadership.

## They Understand Team Dynamics

The pressure of extreme environments can strain team dynamics, making it crucial for extreme leaders to excel in maintaining team cohesion and motivation under highly volatile and emotionally draining conditions.

## They Prepare and Train

Training for extreme leaders often involves rigorous physical and psychological preparation, reflecting the demanding nature of their roles. In contrast, conventional leadership development typically focuses more on strategic thinking and management skills, with less emphasis on physical and emotional endurance.

## They are Alone in Believing Their Initial Vision

Extreme leaders often find themselves alone initially, with no one else understanding or buying into their vision. While others may eventually feel, relate to, and be drawn to the vision, they rarely grasp it at the outset. Some will dispute or even work against it, criticizing and discrediting the leader as a dreamer, much like Christopher Columbus, or as a charlatan or witch, as Joan of Arc was labeled. Despite this, extreme leaders use their initiative to kick off a chain of events.

## Their Dreams Are Deemed Implausible

Extreme leaders' visions are typically grand, seemingly impossible, and beyond most people's imaginations. These ambitious dreams are a powerful source of their influence because they intrigue, ignite the imagination, and inspire. Despite often facing impossible odds, extreme leaders find a way to overcome them.

### They Tap into Dissatisfaction with the *Status Quo*

Extreme leaders often sense the great tension affecting those around them—a tension that people are unwilling or afraid to acknowledge, speak about, or address. Joan of Arc tapped into the frustration of French citizens who feared and hated the English invaders yet were afraid to act until she stepped forward. Similarly, Adolf Hitler rose to power by promising the German people what they longed for, reinstating their pride, and restoring their economy to pre-World War I levels. People's dissatisfaction, like lightning, is a powerful force when a leader can capture and channel it.

### They Find Courage

Extreme leaders summon the courage to confront issues others wish to address but have been unwilling to face alone. Once a leader steps forward, many people will quickly offer support, recognizing that their frustrations will persist without decisive action.

### They Do Not Wait for Someone Else to Step Up

Extreme leaders summon the courage to step forward and lead people in solving their most frustrating problems without waiting for someone to ask or direct them. They do not wait for someone else to act; they take it themselves. Often, extreme leaders admit to experiencing great fear about what they might face, but they always find a higher purpose that drives them to confront challenges despite the dangers.

### They Attract Followers

Extreme leaders embrace the responsibility for the mental and physical well-being of others. When the vision is compelling, and the pain of the current situation is acute, people will step up to follow. Over time, they become loyal and as brave as their leader.

### They Risk All They Own

Just as America's Founding Fathers risked their fame, fortunes, and lives against the British, extreme leaders do the same. Figures like Christopher Columbus, Hernando Cortes, and Joan of Arc gambled everything—money, reputations, and lives. Nothing sharpens focus and energizes a person like having everything on the line. In many ways, this all-or-nothing commitment is a powerful energy source for extreme leaders.

### Their Means Are Greater

Extreme leaders often start with few physical resources, limited money, and minimal supporters, relying on their personality to gather what they need. They succeeded in raising resources due to their charisma, persuasive skills, earnestness, and what Mark Twain referred to as "a transparency of sincerity" (Twain, 1899).

These leaders frequently do not hold formal leadership positions when their vision arises. Instead, their leadership role emerges from their passion, commitment, and persistence. Additionally, extreme leaders typically exhibit extraordinary willpower and patience.

> Christopher Columbus pursued Queen Isabella's support for 12 years. Joan of Arc had to present her request and plead her case to the governor of Vaucouleurs three times over a year to secure an escort of men-at-arms to see the king.

### They Are More Deeply Moved by Trying Circumstances

Extreme leaders recognize the crisis around them and deeply understand that someone must address it. This internal drive helps them remain composed in emergencies, stand firm when others flee, and maintain confidence in their ability to survive regardless of the circumstances.

### They Are Prepared to Give Their Life for Their Vision

Martin Luther King is one of the finest examples of an extreme leader. He recognized injustices that needed to be addressed, which others saw but were doing little about. He crafted a vision for equality and pursued it with unrelenting passion, ultimately sacrificing his life for his cause.

> Extreme leaders often anticipate that they may give their lives for their cause. Martin Luther King knew, Joan of Arc knew, Mahatma Gandhi knew, and even Adolf Hitler, in his own destructive pursuit, understood the potential for his life to be forfeited.

Not all extreme leaders die as a result of their quest. For example, Giuliani, Patton, and Cortes achieved their Big Visions without paying the ultimate price. While death is not a prerequisite for extreme leaders to achieve their goals, it can be a consequence.

### They Maintain Positivity Despite Setbacks

When faced with unexpected challenges, extreme leaders rearrange priorities, set new goals, and initiate new actions to maintain progress. They see opportunities others overlook and find creative solutions to seemingly impossible problems.

### They Face Danger with Unique Confidence

Extreme leaders sometimes face extreme danger, becoming trapped and battling nature's elements or an adversary to escape. Sometimes, they are imprisoned or even killed, as with Joan of Arc, Martin Luther King, and Nelson Mandela. However, they confront these trials with grace and, at times, resignation toward the inevitable—an inevitability they had anticipated and contemplated. Sometimes, as with climbing Mount Everest, the reward may not seem worth the risk after the first ascent. The initial achievement, while monumental, often highlights the extreme dangers and challenges involved, leading some to question whether the potential rewards justify the continued risk. One of every twenty climbers who reached the summit died.

### They Break Unjust Rules

In many cases, extreme leaders find they must violate laws or societal and political rules because these regulations are unjust or hinder the accomplishment of their visions. Extreme leaders may break the rules peaceably, as Martin Luther King and Mahatma Gandhi did, or may resort to violence or force, when necessary, as Nelson Mandela did before his imprisonment and Cortes when he covertly moved his ships out of the harbor. Violence is a measure of last resort for extreme leaders with noble purposes. In contrast, those with evil intentions use violence to seize power and perpetrate evil deeds.

### They Draw Their Authority from a Higher Authority

In some cases, extreme leaders, driven by religious fervor, believe that a higher calling motivates their actions. An inner force propels these leaders, and they may also receive support from spiritual enlightenment or guidance.

### They Leave a Lasting Impact

Attempting a Big Vision, whether successful or not, extreme leaders leave a lasting impact regardless of the outcome. Adolf Hitler left a horrifying legacy by orchestrating the genocide of millions of Jews. In contrast, Joan of Arc's efforts to liberate France from English rule had a profoundly positive and enduring effect. Similarly, the mission to walk on the moon created an everlasting mark in history.

## They Are Clear-Headed and Mentally Strong

Extreme leaders may vary in physical toughness, but they always clearly understand what they want and what needs to be done to achieve it. They are mentally sharp and highly persuasive, convincing others to believe in their vision and viewpoint.

## They Excel in Unstable, High-Expectations Environments

Extreme leaders may only be as secure as their last victory. Napoleon won and lost significant battles, was exiled, returned to lead more campaigns, and ultimately faced a substantial defeat in Russia, leading to his final exile by his supporters.

For extreme leaders, the maintenance of high standards means that one defeat may be tolerated, but multiple losses can quickly erode confidence and cause supporters to drift away or turn against the leader. This compounds the leader's problems, creating a potential downward spiral.

## They Are Sensitive to Others

Extreme leaders build, discover, conquer, invent, create, and achieve new vistas. They also soothe, listen, heal, touch, and bless, as Mother Teresa did consistently, and Joan of Arc frequently did. Even Cortes had moments of compassion, such as when he forgave a mutineer for a first offense, though he later executed him for a second. Extreme leaders maintain humility at all times.

"Humility must always be the portion of any man who receives acclaim earned in the blood of his followers and the sacrifices of his friends." — Dwight D. Eisenhower.

## They Seek Support from the Wealthy and Powerful

Most, if not all, extreme leaders have received help from wealthier or more powerful individuals but did not let that support influence or divert them from their ideals. Leaders who make decisions primarily based on personal gain do not fit my definition of an extreme leader, as their purpose has no nobility.

Dr. Cohen's seven steps for leaders taking charge in extreme circumstances are invaluable. He believes a leader in such situations will do the following:

1. **Establish the Objective**: Clearly articulate why the effort is essential.

2. **Communicate Effectively**: Speak calmly, loudly, clearly, and colorfully to ensure the message is understood.

3. **Act Boldly but Not Recklessly**: Do not wait for more information or risk delays, changing circumstances, lost opportunities, competitors getting ahead, or followers losing heart.

4. **Decide and Communicate Immediately**: Once decisions are made, promptly inform everyone who needs to know.

5. **Dominate the Situation**: Take decisive actions to gain and maintain control, staying ahead of opponents and events. Do not limit expectations of what followers can achieve based on their self-imposed limitations.

6. **Lead by Example**: Be willing to do whatever you ask of your followers and prioritize their needs before your own.

7. **Hire Strong and Fire Weak**: Replace those who consistently perform poorly and cannot be redeemed, provide private reprimands to those likely to respond to positive feedback, and publicly praise those who perform well.

---

**Key Attributes of Extreme Leadership**

1. **Clarity in Chaos:** Mastering decision-making under pressure.

2. **Resilience and Grit:** Building mental and emotional toughness.

3. **Rapid Adaptability:** Demonstrating flexibility in strategies and tactics.

4. **Contingency Planning:** Anticipating and preparing for the unexpected.

5. **Quick Thinking:** Adapting and responding swiftly in dynamic situations.

6. **Performance Under Stress:** Excelling in executing tasks effectively amidst stress.

---

The above characteristics can be found among good leaders who rise to the challenge when confronted with extreme circumstances. However, these traits are present to a greater degree in those who are inherently extreme leaders. Additionally, extreme leaders possess other qualities that further distinguish them.

Chapter 6

# ADDITIONAL QUALITIES OF AN EXTREME LEADER

Extreme leaders are extraordinary—not necessarily as individuals but as visionaries due to their exceptional focus and commitment. In addition to the qualities previously described, they may or may not also possess traits such as the following:

1. **Innovative Thinking**: They continually seek out new solutions and approaches.

2. **Inspirational Presence**: They have a remarkable ability to motivate and inspire others.

3. **Unwavering Integrity**: They maintain a solid ethical foundation, even under pressure.

4. **Strategic Vision**: They can see the bigger picture and plan long-term strategies.

5. **Courageous Action**: They are willing to take bold risks for their vision.

Having explored the characteristics and essential qualities of extreme leaders, we will now delve into the specific traits and strategies that empower them to achieve remarkable success and elaborate on some of the characteristics we outlined above.

## A Big Vision

Extreme leaders possess a vision grand enough to inspire others to pursue it alongside them. This vision often resonates deeply with people because it addresses shared frustrations or beliefs. Such a vision typically involves significant risks but is compelling enough to attract followers willing to face these extreme risks. These followers are often highly dedicated to the cause, accepting the risks for little if any, ultimate reward beyond the mission's success. Given the inherent dangers, they are rarely compensated proportionately for their sacrifices and efforts. The vision must also capture the attention of contemporary media. In today's environment, an extreme leader's message must be "breaking news" to effectively reach and mobilize potential followers.

## Perfect Timing

Extreme leaders possess an acute sense of timing, knowing precisely when to act. Victor Hugo, the renowned French poet, playwright, novelist, and essayist, famously stated, "Nothing is more powerful than an idea whose time has come." This insight is particularly relevant to extreme leaders who intuitively recognize the right moment before anyone else. The sentiments of their followers often drive the timing of their actions; when the followers reach a point of having "had enough" and are ready to stand up and fight, which is when the extreme leader steps forth.

A critical skill of extreme leaders is their ability to assess when their followers' will to fight surpasses their opponent's will to resist. This judgment enables them to mobilize their supporters effectively and seize the moment for action.

## Clarity of Purpose

Extreme leaders clearly understand the mission's purpose and the sacrifices necessary to achieve it. There is no ambiguity or confusion in their minds. They feel driven by a predetermined destiny and actively embrace their role, knowing it is their fate to lead. "good" extreme leaders' purpose is always rooted in a constructive and noble cause that benefits many. In contrast, "evil" extreme leaders may be motivated by selfish desires, such as a quest for power, which benefits only themselves or a select few. They may even believe their cause is noble, but society will eventually recognize it as misguided, as seen in Hitler's goal to "purify the race."

Good extreme leaders inspire through their dedication to the greater good, whereas evil extreme leaders manipulate and exploit for personal gain. Understanding this distinction is crucial in recognizing the true nature of leadership and its impact on society.

## Constancy of Purpose

Purpose Clarity also requires purpose constancy for extreme leaders. They must remain undaunted by the risks they face. To them, disappointment is merely a temporary setback, delaying what they consider inevitable. Extreme leaders are not deterred by disappointment but resilient in the face of impediments or delays. They are rarely turned away by overwhelming opposition. When driven away from their goal, they will redouble their efforts and persist.

## Passion

Extreme leaders possess a profound passion for their purpose, which naturally extends to their Big Vision. This fervor ignites the passions of their followers. For example, Cortes's love for gold was intertwined with a fervent desire to form a religious crusade to conquer Mexico and convert its pagan inhabitants. His intense commitment inspired and mobilized his followers, demonstrating how an extreme leader's passion can galvanize others to join their cause.

## Determination

Extreme leaders are undeterred by disappointments, impediments, uncontrollable delays, or overwhelming opposition. When driven off course or away from their goals, they fight back by finding creative alternative solutions and persevering until they achieve their objectives. They are willing to risk their lives, fortunes, and even the success of their enterprise to fulfill their core purpose. These leaders are equally prepared to lay down their lives for their followers, knowing that such a sacrifice will inspire them to complete the mission. Driven more by creed than conduct, extreme leaders often start as naive idealists and evolve into experienced, victorious generals or respected, long-remembered martyrs.

## Preparation

Many extreme leaders prepare themselves for their quests through experience, meditation, practice, reading, advisors, and training. When resources are unavailable, they create them seemingly out of thin air by finding sympathetic supporters or gathering resources along the way. Extreme leaders often achieve vast accomplishments with seemingly inadequate means.

### A Caring Nature

Extreme leaders are generally sensitive individuals who care deeply about each follower and others they encounter. They can also be magnanimous toward the enemy, provided it does not jeopardize the mission. Extreme leaders embody the concept of a romantic enterprise without romanticizing it.

### A Take-Charge Demeanor

Extreme leaders are not shy about taking charge. They often step forward before sponsors, followers, or opponents are ready for them. These leaders quickly become the heart and soul of the endeavor. From the first moment, they know the goal and the general approach to achieving it, even if they do not have all the details worked out. When fighting begins, extreme leaders are fully involved in the thick of the battle, leading the fighters and facing whatever comes their way. They handle all negotiations, intrigues, and correspondence related to the mission, often keeping a diary or writing commentaries as the mission progresses.

Extreme leaders are cautious and calculating in their plans but do not delay unnecessarily. They are known to stroll among their followers during downtimes to gauge their condition and look after their well-being, even standing with the person on the hardest watch. During the pitch of battle, extreme leaders are so enlivened by seeing their plan in action that their eyes have been described as glowing from deep inside. Often, they fully appreciate the mark they are making on history.

### Access to Good Generals

Extreme leaders attract talented individuals and often have a close confidant as their chief of staff. Along the way, they draw others who serve as generals, helping to execute their plans. This ability to attract and mobilize skilled people is crucial for the success of their ambitious visions.

### Power Over the Minds of Others

Extreme leaders excel at winning the hearts of tough men and women and forming close bonds with those nearest to them. This trait is their genius—the ability to unite a diverse group of volunteers or underpaid followers and motivate them to fight under their banner for a cause that initially seems impossible. This remarkable talent enables extreme leaders to build a committed and cohesive team, driving them toward achieving extraordinary goals.

Cortes, the early explorer and invader of Mexico, exemplified an extreme leader. His entourage consisted of individuals driven by the pursuit of gold, fame, or redemption. These men were reckless veterans, vagabonds fleeing justice, former enemies, and various opportunists. Despite their differences in race, language, social class, and interests—factors that typically breed infighting and factions—Cortes unified them under a common purpose. He convinced them to work harmoniously toward a shared, noble principle, granting him total authoritarian leadership.

Cortes eschewed dressing to his station, opting for simplicity, yet maintained an air of dignity. He spoke eloquently, like a poet, without exuding superiority. Charitable to the poor and patient with his men, Cortes treated them as equals, even allowing them to call him by his first name, Hernando. He educated his captains on the land's resources, social organization, and physical capabilities, prioritizing his men's welfare over his own. Cortes enforced strict obedience and severe discipline but never publicly criticized his men. He did not tolerate abuse towards opponents and avoided unnecessary destruction, often turning staunch enemies into allies. He aimed to leave behind a better culture and a higher civilization according to the standards of his time. This leadership style allowed Cortes to transform a diverse and potentially fractious group under his command into a cohesive and motivated force. (Prescott, 1843 vol. II)

Being an extreme leader also has its downsides. Many historic extreme leaders, such as Mahatma Gandhi and Martin Luther King Jr., attracted enemies due to the rapidity and grandeur of their successes, misunderstood motives, accusations of personal aggrandizement, or imagined wrongs. These leaders sometimes faced disdain for their intellectual superiority, cultivated through solid early education, voracious reading, writing elegance, and talents in math, logistics, geography, and other relevant skills. Additionally, simple jealousy of their success or a twisted desire to achieve immortality in the annals of history could fuel animosity toward them.

# Chapter 7

# THE NATURE OF EXTREME SITUATIONS

Unpredictability, high stakes, and an urgent need for swift decisions characterize extreme situations. Due to their minimal margin for error, these high-risk environments demand unique management and response strategies. Effective leadership requires a clear head, rapid adaptability, and the ability to take decisive actions under pressure.

## Extreme Situations Defined

Extreme situations are marked by unpredictability, high stakes, and an urgent need for swift action. They often occur in environments where risks are substantial, and the margin for error is minimal. Grasping these dynamics is essential for managing and responding effectively in such conditions.

## Extreme Situations Characteristics

The hallmark of extreme situations is their unpredictability, which arises from the complexity of the factors at play, making outcomes difficult to forecast. This unpredictability is compounded by the high stakes involved, where the consequences of actions—or inactions—can be profound, potentially resulting in severe loss or damage. The urgent need for a rapid response adds another layer of complexity, as decisions must be made quickly and often with incomplete information.

In such scenarios, effective decision-making requires a blend of experience, intuition, and the ability to remain calm under pressure. Leaders and managers must be adept at quickly assessing situations, prioritizing actions, and mobilizing resources efficiently. It is also crucial to have contingency plans in place and to be adaptable, as initial plans may need rapid adjustments in response to unfolding events.

Communication takes on a pivotal role in extreme situations. Clear, concise, and timely information exchange can significantly influence the effectiveness of the response. Any form of miscommunication or delays can worsen the situation, leading to greater risks and potential failures.

Understanding the dynamics of extreme situations and preparing for such scenarios through training and simulations is a proactive approach that can significantly enhance the ability to manage and respond effectively. This preparation can be the deciding factor between successful navigation through a crisis and catastrophic outcomes.


### Examples of Extreme Situations

Additionally, extreme situations in these fields often require high levels of teamwork, coordination, and effective communication, as well as the ability to adapt to rapidly changing conditions.

## Challenges Unique to Extreme Situations

One of the primary challenges in extreme situations is managing the intense pressure and stress that accompany high stakes and urgent timelines. This stress can impair cognitive functions, making clear thinking and effective decision-making more difficult—precisely when they are most needed. The inherent unpredictability and volatility of these situations add another layer of complexity. Leaders and teams must be adaptable and ready to pivot their strategies as new information becomes available.

Furthermore, the limited time available for decision-making heightens these challenges. Decision-makers are forced to depend on their training, instincts, and available data to swiftly make the best possible choices. Successfully managing these scenarios necessitates not only technical skills, but also emotional resilience and the ability to maintain composure under pressure, underscoring the importance of these resources in high-pressure situations.

Chapter 8

# LIMITATIONS OF TRADITIONAL LEADERSHIP PRINCIPLES

Traditional leadership principles often fall short in extreme situations, revealing critical limitations in risk management, decision-making processes, and hierarchical structures.

### Inadequate Risk Management in Extreme Situations

In traditional settings, leadership often emphasizes calculated risks, where uncertainties can be predicted, measured, and mitigated. However, extreme situations present a different paradigm, characterized by uncertainties that are difficult to predict and measure. Traditional risk assessments may address predictable challenges but often fall short during crises where dynamics change rapidly and unexpectedly. These assessments may fail to consider the immediacy and interconnectivity of various risk factors, leading to catastrophic oversight.

### The Pitfalls of Slow Decision-Making Processes

Conventional decision-making processes are typically methodical and deliberate, involving extensive information gathering and consensus-building. While this approach suits routine business decisions, it becomes impractical in the time-sensitive situations typical of extreme scenarios. In corporate environments, the luxury of time allows for detailed analysis and thorough discussion. In contrast, emergency services must make rapid decisions with limited information to respond effectively to immediate threats.

### Overreliance on Hierarchical Structures

Traditional leadership models frequently rely on rigid hierarchical structures that dictate decisions from the top down. While this can ensure coordinated efforts under normal conditions, it can become a liability in extreme situations where quick, autonomous decisions are needed at the front lines. In military operations, for instance, the ability of lower-ranking personnel to make field decisions can be critical. Rigid hierarchies can hinder such autonomy, potentially delaying actions crucial for mission success or survival.

### Adjusting to Different Stakes and Stress Levels

In typical workplace environments, standard motivational strategies are often sufficient to engage team members. These strategies may include goal setting, recognition, and rewards. However, in extreme environments where team members face high levels of stress or personal danger, these traditional approaches may fall short. Under extreme conditions, the motivations of individuals often shift toward more intrinsic and survival-focused needs. Recognizing this shift is crucial for leaders, who must develop motivational strategies that resonate on a more fundamental, survival-oriented level. These strategies should emphasize resilience, the significance of each team member's role in collective safety, and the direct impact of their actions on the outcome of high-stress situations.

### Addressing Cultural Misalignment

Traditional corporate cultures and leadership training programs are designed to support stability, efficiency, and growth within predictable, controlled environments. However, these frameworks may not align well with the values and immediate needs of teams operating in extreme environments, such as military units or disaster response teams. In such settings, the cultural imperatives focus on rapid response, adaptability, and cohesion under pressure rather than on long-term strategic planning and individual competition. The necessity for cultural adaptation becomes apparent when comparing military and corporate leadership frameworks. This calls for a revaluation of the underlying values and training methodologies to develop adaptive leadership models better suited to the demands and stresses of extreme situations.

Chapter 9

# CHARACTERISTICS IN THE MIDST OF THE FRAY

Extreme leaders display determination and resourcefulness while organizing to pursue their Big Vision. However, in the midst of conflict, other characteristics emerge. It is one thing to be determined while rallying supporters and convincing key allies to join your quest. However, displaying that same determination in the heat of battle elevates it to an entirely higher level.

When the battle begins, the extreme leader must often lead from the front, inspiring followers to risk their lives and livelihoods. Followers will match the leader's bravery and commitment, making it essential for the leader to exhibit supreme confidence and resilience in extreme environments. An extreme leader must stand their ground longer than others, maintaining composure and strategic thinking during crises.

Survival is crucial to loyalty; followers commit to leaders who lead successfully through danger. Loyalty may wane if a leader is lost, but extreme leaders risk their lives for the mission, knowing the stakes.

Additionally, extreme leaders must excel in handling dissent. History shows that a "benevolent dictator" approach—being firm yet forgiving—can be effective. Every leader will face challenges to their authority, and both followers and opponents need to see that the leader is willing to take extreme measures when necessary. Otherwise, provocations may escalate, with people exploiting perceived weaknesses until the leader reacts decisively or withdraws.

In the case study below on Dr. Elisha Kane, captain of the Advance, you will see a core strategy for dealing with dissent that cannot be suppressed by force. Kane faced a crisis when the ship became frozen in ice, and some of his men wanted to attempt a 1,000-mile trek in the dead of winter. Bound by maritime rules, he could not compel the men to stay. So, he devised the following approach:

1. He laid out a new plan for those who would stay.

2. He gave the dissenters his blessing to leave and a more than fair share of the remaining stores.

3. He got those who would remain to buy into his plan.

4. He saw the dissenters off with the promise to take them back with a "brother's welcome" should they return.

5. As soon as the dissenters were gone, he immediately got those who remained to work on the new plan.

This is a basic plan for dealing with dissention in an extreme leadership situation.

## How Extreme Leaders Deal with Setbacks

Extreme leaders face setbacks with resilience and flexibility. When confronted with multiple setbacks, they do not give up. Even when thrashed, beaten, and forced to retreat, extreme leaders find ways to recover and attack again. They are also intelligent, innovative, and strategic; rather than wasting resources foolishly, they may retreat under the protection of a rear guard to safer ground to regroup, reenergize, and prepare for future battles.

Extreme leaders are adept at adapting to changing environments and circumstances. They excel at finding creative solutions to outsmart opponents and overcome challenges. Their willingness to make unpopular decisions and their ability to see options others overlook inspire their followers to adjust and thrive in changing circumstances. This adaptability is a key factor in their success.

## Extreme Leaders Have Extreme Resolve

Extreme leaders are resolute in driving toward their objectives, never giving up regardless of the risks, dangers, or obstacles. This unwavering resolve, which may seem superhuman to followers, is simply the manifestation of extraordinary determination. Extreme leaders, with their relentless persistence, push the limits of human endurance and withstand extreme conditions through sheer willpower. Moreover, they possess the unique ability to see a way out of a crisis, even when it appears impossible to others.

## Extreme Leaders Innately Understand Strategy

Extreme leaders cultivate an early interest in strategy, developing an innate understanding of its principles. This deep comprehension encompasses maintaining a continuous, balanced offense with rapid movement, improvisation, and the element of surprise. Their strategic insight allows them to adapt and excel in dynamic environments, turning theoretical principles into practical, real-world applications. (Patton, 2024, *Strategies and Tactics*).

## Extreme Leaders Judge Character and Talent Keenly

Extreme leaders have a unique ability to identify the best followers. They don't just accept anyone willing to follow; they choose individuals who bring critical experience, positive attitudes, and potential for high loyalty and are naturally inclined to follow. While they may seek resources from higher authorities, their true power comes from their followers. These followers often need very little that money can buy, but their unwavering support is invaluable, fuelling the leader's effectiveness and success.

## Extreme Leaders Master a Broad Range of Topics

Extreme leaders acquire essential knowledge in crucial areas such as strategy, science, and psychology through extensive reading, discussion, and observation. They leverage this knowledge to exploit opponents' weaknesses, such as indecisiveness, sluggishness, lack of willpower, physical weakness, or disorganization.

## Summary of Extreme Leader Characteristics

Extreme leaders excel by adapting traditional principles to thrive in unpredictable, high-stakes environments. They swiftly adjust their risk management strategies to address rapidly changing threats, making quick, informed decisions even under intense pressure. By decentralizing decision-making, they empower their teams to act autonomously and effectively in critical moments. Understanding the unique stressors of extreme situations, these leaders implement motivational strategies that resonate on a fundamental level, ensuring team cohesion and resilience. Additionally, they bridge cultural gaps by fostering a culture of rapid response and adaptability, aligning their leadership approach with the demands of extreme environments to achieve remarkable success. Many of the traits needed for a leader to become an extreme leader can be learned.

Chapter 10

# HOW TO DEVELOP EXTREME LEADER TRAITS

The preceding chapters describe the characteristics and qualities extreme leaders need to succeed. This chapter will discuss how these traits are developed. An extreme leader has the following.

### Strength of Mind Under Duress

Training under realistic simulations of expected environments is the best preparation for maintaining mental strength during crises. This practice is why police academy training includes live-fire exercises in a mock village with randomly appearing targets that represent both threats and innocent people. I once experienced this training, encountering a scenario where a large woman holding a baby pointed a gun at me. Making the right decision in such ambiguous situations is neither easy nor clear-cut. (I will not say what I did).

Experience builds self-confidence, character, conditioning, and the ability to make quicker decisions. While simulated training exercises cannot guarantee courage in the face of fear, they significantly reduce the likelihood of a leader's courage failing during real crises. Exposure to potential scenarios enhances a leader's readiness and resilience. Additionally, a strong will can be developed through meditation alongside rigorous training.

> Martial arts teachers often recount the story of samurai who were fearless in battle because they convinced themselves beforehand that they were already dead. By accepting their death, they had nothing left to fear. Whether or not this story is true, the concept illustrates the profound spiritual conviction required to embrace such a mindset. If achieved, this perspective could enable one to act with complete fearlessness.

Strength of mind requires self-control under duress, which primarily comes from experience or, if not from experience, then from a firm conviction of purpose. Simulated training exercises and real-life experience aim to develop healthy instincts that can be relied upon during crises. While a strong conviction of purpose may drive extreme leaders to push through problems, experience can reveal alternative actions that might avoid unnecessary difficulties, such as negotiating instead of confronting.

The strength of personal will, which is fully embodied by extreme leaders, is their source of resilience under pressure. When leaders are resolute in their belief in their cause, they can overcome many obstacles and persevere through significant challenges.

### An Unshakable Firmness

Extreme leaders possess stable views rooted in fundamental beliefs and principles, shaped by deep reflection and extensive experience applying those principles. These principles are tested with followers to ensure they resonate and are further honed through challenges against opposition. Such firmness is often tempered with a delicate balance of qualities like caring, fairness, dependability, and trustworthiness, making it both strong and harmonious.

Opponents can sense this unwavering firmness, which in itself can be intimidating. Less committed adversaries often shrink before fiercely committed attackers, recognizing the resolve and determination of extreme leaders.

### Quick and Sound Decision-Making

Extreme leaders, grounded in solid beliefs and principles, can make quicker and sounder decisions. Such decisions are more valuable than extensive studies, unnecessary risk-taking, or time spent correcting mistakes. Quick decision-making can be practiced in all aspects of life. Parents can start teaching children by offering choices and encouraging them to decide. Over time, frequent decision-making in varied circumstances helps children become adept decision-makers. Adults can enhance their decision-making skills by increasing their risk tolerance and consciously making quicker decisions.

Understanding the consequences of a wrong decision is critical to making quick decisions. The larger the consequences or the more uncertain the outcomes, the more time and care should be taken. Quick decisions over minor matters carry little risk and require minimal information. Conversely, decisions with significant ramifications should be delayed as long as practical to gather as much information as possible. Effective quick decision-making involves rapid risk assessment, honed through experience.

## Extreme Leaders' Mastery of Opponent Knowledge

Extreme leaders strive to learn everything about their opponents—past, present, and future. They are well-read, not just about their adversaries but across various subjects. They gather intelligence through various means, including building spy networks, to understand an opponent's willpower, genius, decisiveness, endurance, creativity, and leadership under pressure. The key is to know your opponent inside and out.

A famous example of this is depicted in the movie "Patton," where George C. Scott, portraying General George S. Patton, exclaims, "Rommel, I read your book," while observing a victorious tank battle against German Field Marshal Erwin "The Desert Fox" Rommel. This line signifies that Patton had not only studied Rommel's tactics but had also gained insight into his thinking and strategy, allowing him to anticipate Rommel's moves in battle.

## Knowledge of Their Followers

Extreme leaders deeply understand their followers' concerns, virtues, defects, and alignment with the leader's vision and purpose. They are keenly aware of their followers' mindsets, ideals, biases, and personal characteristics such as honor, honesty, trustworthiness, and ethics. Leaders can better align their purposes and missions by knowing their followers well. Accurate judgment of people, resources, and capabilities requires extreme leaders to have experience, knowledge, and strong will.

## Creativity

One of the more challenging traits for aspiring extreme leaders to develop is creativity. The ability to devise new solutions to old problems or entirely novel ideas is not easily acquired but can be learned through practice. The first step is to be prepared, embodying the Boy Scout motto. Preparation involves anticipating needs and potential scenarios during crises. This proactive thinking, including "what if" scenarios, equips a leader with a portfolio of solutions and a framework for generating alternatives when the initial plan falls short.

## Endurance

Extreme leaders possess exceptional endurance and can withstand prolonged efforts and challenges. This endurance stems from thorough preparation and physical conditioning—being trained, fit, and mentally ready for the demands they face.

## The Boy Scout Oath and Staunchness

The Boy Scout oath—to be trustworthy, loyal, friendly, courteous, kind, obedient, cheerful, thrifty, brave, clean, and reverent—holds significant truth and value. These principles form the foundation of healthy relationships and personal integrity. Practicing these values from a young age contributes to personal development and better character. Young men who achieve the rank of Eagle Scout often succeed because they embody these principles and have been trained to be goal-oriented and disciplined leaders.

Extreme leaders also possess staunchness and the ability

to recover from severe physical or strategic setbacks. This resilience is rooted in thorough preparation, experience, and training, enabling leaders to withstand and bounce back from significant challenges.

### Makes Tough or Unpopular Decisions

Extreme leaders are adept at eliminating or reallocating essential resources when faced with dissent, poor performance, or disruptions that distract from their objectives. Making tough decisions comes naturally to them, driven by their intense focus on the Big Vision. These leaders prioritize the ends over the means, guided by a clear purpose as a foundation for their decision-making. This clarity allows them to act decisively and maintain alignment with their ultimate goals.

### Unwavering Bravery

Extreme leaders never waver when others count on them. They fear being seen as cowards just as much as any follower does. Followers trust their leaders and strive to please them, especially when they believe staying behind is more dangerous than moving forward. Therefore, leaders must demonstrate bravery, which stems from their mental mindset—acting despite fear. This courage inspires confidence and motivates followers to advance together.

## The Right Example

The behavior of extreme leaders is more crucial than that of their followers. Leaders must balance firm direction with personal enthusiasm for shared goals, setting an example that inspires and motivates their team.

## Effective Communication of Extreme Leaders

Extreme leaders' communications promote unity of purpose, characterized by clarity, focus, and vivid language that includes action verbs and stated goals. They also listen closely to their followers to keep them aligned and make necessary adjustments based on feedback. Their communication aims to gather and share information, provide support, build teams, encourage collaboration, offer direction, allocate resources, recognize efforts, and reinforce the rationale behind decisions, the mission, and the Big Vision.

# Chapter 11

# HOW TO BUILD AN EXTREME TEAM

In high-stakes projects, the difference between success and failure often hinges on the caliber of the team. This chapter delves into the art and science of assembling an "Extreme Team," equipped to tackle the most demanding challenges with unparalleled synergy and expertise.

### Recruitment, Selection, and Development

Identifying potential leaders for high-pressure roles requires a thorough process. Here is a guide to selecting the best candidates, focusing on three crucial traits: decisiveness, resilience and emotional stability, and communication skills. These traits support effective leadership in crises, enabling leaders to guide their teams through challenges confidently and clearly.

### Recruitment Techniques:

- **Targeted Sourcing:** Focus on industries or roles demanding high resilience and stress management, such as military veterans, emergency service personnel, or individuals in high-stakes financial or legal positions.

- **Professional Networking:** Utilize professional networking platforms and industry-specific events to identify candidates with proven leadership during crises.

**Hiring and Evaluation Process:**

**Behavioral Interview Questions:** Ask targeted questions that prompt candidates to describe specific instances where they had to make rapid decisions. This approach reveals their thought process, confidence, and speed in decision-making.

1. **Simulations and Role-Playing**: Simulate high-pressure scenarios requiring quick decision-making to see how candidates perform in real-time. Role-playing exercises or virtual simulations can help assess a candidate's decisiveness and effectiveness under pressure.

2. **Psychometric Tests**: Use assessments to measure traits associated with decisiveness, such as risk tolerance, stress resilience, and problem-solving speed. These tests provide a quantifiable measure of a candidate's ability to make effective decisions under pressure.

3. **Psychological Assessments**: Utilize standardized tests to measure emotional intelligence, stress tolerance, and resilience. These tests offer insights into how individuals perceive and manage their emotions and the emotions of others.

4. **Case Studies**: Present candidates with real or hypothetical case studies relevant to your industry, asking

them to devise and justify their action plans. This exercise assesses their analytical skills and decisiveness.

5. **Reference Checks**: Speak with previous employers or colleagues to gain insights into how the candidate handled stress and adversity in past roles. Focus on the candidate's emotional responses to challenging situations and ability to bounce back from hardships.

6. **For Internal Candidates**: Use 360-degree feedback from colleagues, subordinates, and supervisors, and review performance histories. Past performance in roles requiring quick decision-making can be telling.

7. **Group Interviews and Team Discussions**: Observe candidates in group interviews or discussion settings to evaluate their communication effectiveness, listening skills, and ability to advance conversations constructively.

8. **Situational Questions for Communication**: Present hypothetical scenarios and ask candidates how they would communicate in those situations. Request samples of written work (reports, emails, proposals) to assess clarity, brevity, and impact. Have candidates deliver a presentation on a relevant topic to evaluate their ability to convey information clearly and engage an audience effectively.

By integrating these strategies, you can build a resilient, capable team ready to excel in high-pressure environments.

Once the right candidates are hired and recruited the focus shifts to development, ensuring these individuals are equipped with the skills and resilience needed to thrive in high-pressure environments.

## Development Processes

Equipping team members with the necessary skills and resilience through targeted development programs is crucial for ensuring they can effectively handle the demands of high-pressure environments.

1. **Training and Development**: Specialized training programs for extreme scenarios:

2. **Wilderness Emergency Medical Technician (WEMT) Training**: This program is tailored for those operating in remote and extreme environments, such as search and rescue teams, firefighters, and adventure guides. It combines traditional EMT training with wilderness medicine, focusing on skills needed to provide care when access to medical facilities and resources is limited. Trainees learn advanced first aid, evacuation techniques, and long-term care strategies outside typical clinical settings.

3. **High-Risk Operational Training (HROT)**: Designed for military, law enforcement, and private security personnel, this training prepares operatives to handle high-stakes scenarios such as hostage rescues, counter-terrorism operations, and other critical incidents. HROT covers tactical firearms training, close-quarters combat, crisis negotiation, and decision-making under fire, emphasizing mental resilience, leadership, and operational planning under extreme pressure.

4. **Fostering Team Cohesion**: Techniques to build trust and dependability:

5. **Trust Building and Team Cohesion Workshops**: These workshops are designed to improve internal trust and reliability among team members. Through

interactive exercises, role-playing, and facilitated discussions, participants learn about the dynamics of trust, effective communication, conflict resolution, and mutual support in achieving common goals. Activities often challenge teams to work through scenarios requiring collaboration to succeed, reinforcing practical lessons.

6. **Leadership Development Programs**: These programs are targeted at emerging and existing leaders and emphasize developing traits and skills that enhance trust and dependability among leaders and their teams. Components include emotional intelligence training, ethical leadership, and accountability strategies. Participants learn to be transparent, make consistent and fair decisions, and communicate openly—crucial for building trust. Additionally, these programs teach how to inspire teams, creating an environment where dependability is expected and rewarded.

By integrating these strategies, you can build a resilient, capable team ready to excel in high-pressure environments. By integrating these strategies, you can build a resilient, capable team ready to excel in high-pressure environments.

### Becoming an Extreme Leader

Can anyone become an extreme leader? Not just anyone, but those who genuinely believe they can are already on the right path. Can anyone become a better leader? Absolutely. The first step for any leader is to choose a good or evil purpose.

Chapter 12

# GOOD VERSUS EVIL EXTREME LEADERS

Extreme leaders with good intentions, positive attitudes, and constructive motives will be supported and revered. In contrast, those with evil intentions, negative attitudes, and destructive motives must be promptly confronted and eliminated. However, the perception of good or evil can depend on the observer's point of view. For instance, after Adolf Hitler's defeat in World War II, some Germans still believed in his goals and purpose. Even today, groups like Neo-Nazis regard him as admirable or a great man. Despite this, the overwhelming consensus is that Hitler's actions were evil, immoral, and rooted in a perverted mindset.

Evil leaders never see their objectives as evil. They create justifications and rationales that, in their minds, attach a logical and noble purpose to their cause. Therefore, any extreme leader must carefully and thoroughly examine the validity of their so-called noble cause. They must ask themselves, "How will my Big Vision be viewed in the light of history?"

Judging extreme leaders based on the outcomes of their actions allows us to classify them accurately. For example, while Mahatma Gandhi is revered by many, there are likely some Muslims who may disagree due to his Hindu background, despite his support for all religions. Mother Teresa, a Catholic nun, is widely considered an extreme leader with few dissenters. Extreme leaders can be judged by their impact on history and mental state.

### What Was Their Mental State?

Some extreme leaders with evil purposes have lived under delusions of grandeur, paranoia, or grand theories— examples include Karl Marx, Adolf Hitler, Hernando Cortes, and, as some may argue, Donald Trump. However, extreme leaders can be judged by their results, impact on society, and the classes and numbers of people they have helped.

### How Did They Conduct Themselves During the Fray?

Extreme leaders can be judged by the example they set. Did they risk their own lives as readily as they risked the lives of others? Did they lead from the front in times of danger? Did they relinquish power gracefully when their time had passed?

### How Were They Viewed by Followers in the End?

Extreme leaders can be judged by the loyalty and reactions of their followers. Did their followers lionize them both during and after their campaigns? Did their followers eventually turn against them due to shifting public opinion? How did the leaders respond to such adulation? Did they revel in it or demonstrate true humility by shying away from it?

## Tests for Identifying Extreme Leaders as Evil

Extreme leaders can be evaluated based on the morality of their purpose and the behavior of their followers:

- Did followers naturally support their leader, or were they coerced or unthinkingly following mob mentality?

- What level of self-esteem did the followers have? Were they low-esteem individuals easily swayed or confused?

- Did the leaders use propaganda to manipulate their followers?

- Were initiation processes and brainwashing methods used to create "cult-like" commitment?

- Did the leaders try to isolate their followers from the rest of the world?

- Did the leaders have an agenda supported by only a narrow group of followers who did not think for themselves and followed unthinkingly based on the leaders' interpretations (or misinterpretations) of beliefs that mainstream holders of those beliefs did not support?

- Were communications from the leaders restricted to a perceived need-to-know basis rather than being openly shared with all?

- Did the leader instill fear as a means to acquire and retain power?

- Was blind devotion to the leader expected or demanded?

- Were followers treated as subservient pawns or servants?

- Were ideas openly discussed, or were specific topics forbidden?

- Were the leaders trying to impose a radically new order, placing themselves as the supreme leader and conqueror?

- Did the leaders resort to force without attempting to negotiate?

- Were the rights of those forced to change their behaviors respected in governing them in the future?

By examining these aspects, we can assess extreme leaders' morality and integrity and impact on their followers and society.

### Examples of Groups Led by Extreme Leaders

Groups led by extreme leaders fitting the above descriptions include radical terrorists who attempted to further the Islamic religion through force and violence, such as those led by Osama Bin Laden; Nazi Youth Camps organized by Adolf Hitler before and during World War II; the Conquistadors led by Hernando Cortes; Jim Jones and his People's Temple in Jonestown, Guyana; Napoleon and his French troops who attempted to conquer Russia; and Genghis Khan, who overran the largest land mass in history.

One key measure to judge these leaders is: "What was the impact on the people in the affected areas?" Did their actions leave the people better or worse off? This fundamental question helps assess the true legacy and morality of their leadership.

## Tests for Identifying Good Extreme Leaders

Good extreme leaders can be judged by the honorability of their noble cause. Is their cause aimed at righting a wrong rather than building power? Do they serve their followers instead of demanding service from them? Are they concerned about their rewards, or do they prioritize the welfare of their people? Does society increasingly embrace the leader's Big Vision as its accomplishment nears? Are their communications proactive, supportive, and encouraging rather than deceptive? Do their governing methods allow others to have a voice? Is progress made through evolutionary rather than revolutionary means? Does the leader sacrifice themselves for the cause during its accomplishment or aftermath, and does their death evoke societal sympathy?

Examples of good leaders include Nelson Mandela, who reversed one of the most radical and racially biased societies through perseverance, persuasion, and personal sacrifice. Other examples are Christopher Columbus, Martin Luther King Jr., Mahatma Gandhi, Mother Teresa, Dr. Elisha Kane, and the first men on the moon: Buzz Aldrin, Neil Armstrong, and Michael Collins. Additionally, Winston Churchill, Dwight D. Eisenhower, and Antarctic explorer Ernest Shackleton are often considered exemplary leaders.

The judgment of whether extreme leaders' purposes were good, or evil has evolved, as has the understanding of what constitutes extreme leadership.

## Tests for Identifying Bad Extreme Leaders

Bad extreme leaders can be judged by the destructiveness of their cause. Is their cause aimed at building personal power rather than righting a wrong? Do they demand service and loyalty from their followers without offering support? Are they primarily concerned with their rewards rather than the welfare of their people? Does society increasingly reject the leader's vision as it becomes clearer? Are their communications filled with lies and manipulation instead of being supportive and encouraging? Do their governing methods suppress the voices of others? Is progress made through violent revolution rather than gradual evolution? Does the leader seek to maintain power at all costs, even if it means the destruction of their society?

Examples of bad leaders include Adolf Hitler, whose actions led to unimaginable suffering and death during World War II; Osama Bin Laden, who orchestrated terrorist attacks causing widespread destruction and loss of life; Jim Jones, who led his followers to a tragic mass suicide in Jonestown; and Hernando Cortes, whose conquests led to the brutal subjugation and exploitation of indigenous peoples. Other examples are Napoleon Bonaparte, who caused widespread devastation in his quest for power, and Genghis Khan, whose conquests were marked by extreme violence and destruction.

Judging bad extreme leaders involves evaluating their negative impact on society, the coercive methods they use to maintain power, and the extent to which their actions are self-serving and destructive. As with good leaders, the perception of their leadership can change over time, but the lasting harm they inflict defines their legacy.

## Chapter 13

# COMMUNICATION IN CRISIS

Ensuring clarity and precision in critical moments is vital, especially in high-pressure environments where quick and effective communication can significantly impact outcomes. The key is to be prepared! Here are some strategies to achieve clarity and precision.

**Clear Communication Protocols**

1. **Establish Clear Policies and Guidelines:** Develop comprehensive policies and guidelines, secure approval, and train everyone on these foundational (must-do) and recommended (should-do) practices.

2. **Standardized Protocols:** Use clear, standardized communication practices known and practiced by all team members. This approach includes using simple language, avoiding jargon, and employing universally understood codes or signals.

3. **Checklists and Scripts:** Implement checklists and scripts for critical processes to ensure that everything is completed, and all communication is precise.

## Training and Simulations

1. **Regular Drills:** Conduct regular training sessions and simulation drills to prepare for emergencies. This practice ensures everyone understands their role and the communication protocols under stress.

2. **Role-playing Scenarios:** Use role-playing exercises to practice crisis communication and decision-making, helping team members refine their ability to communicate clearly and effectively.

## Example:

> I worked for a major airline for 12 years, leading a small ad hoc team for a year. Our task was gathering all available information about passengers during an airline emergency—a complex and rare responsibility critical in urgent situations. We conducted monthly drills and role-plays, eventually leading to complaints from the team about the frequency, but I insisted on rigorous practice. Then, one day, I got "the call" about a suspected hijacking. Thanks to our consistent training, my team executed their duties flawlessly. When I delivered the results to the central situation room, I arrived just as the information was demanded. A minute's delay would have meant failure, but our frequent drills ensured our success.

## Technology and Tools

1. **Real-time Communication Tools**: Utilize tools that facilitate instant and reliable communication, such as secure radio systems, instant messaging apps tailored for professional use, and video conferencing tools that work under various conditions.

2.  **Information Management Systems**: Deploy systems that can deliver real-time data and updates to all relevant parties, ensuring everyone can access the same accurate information.

3.  **Artificial Intelligence Systems (AI)**: Develop and train AI systems to support operations and provide timely information.

## Feedback Loops

1.  **Immediate Feedback Mechanisms**: Incorporate quick feedback loops in communication practices to ensure messages are understood as intended and allow for immediate clarification if needed.

2.  **Debriefings**: Conduct regular debriefings after actions or decisions to discuss what was effective and what could be improved in communication.

## Structured Communication Techniques

1.  **SBAR (Situation, Background, Assessment, Recommendation)**: Encourage using structured communication models like SBAR to enhance clarity and efficiency in conveying critical information.

2.  **ICED (Introduce, Concern, Explain, Decision)**: Use frameworks like ICED to ensure clarity when introducing a problem, expressing concerns, explaining the situation, and describing the decision or action needed.

## Minimize Information Overload

1. **Prioritize Information**: Ensure that communication is clear and concise, delivering only the most critical information to avoid overload.

2. **Visual Aids: Use diagrams**, flowcharts, or infographics to convey complex information quickly and clearly.

## Cultural and Emotional Awareness

1. **Tailor Communication Style**: Understand the cultural and emotional contexts of team members. Adjusting the communication style to fit the audience can enhance understanding and responsiveness.

2. **Emotional Regulation Training**: Train leaders and team members in emotional regulation to maintain clarity and precision in communication under pressure.

## Crisis Communication Plan (CCP):

1. **Prepared Statements and Templates**: Have pre-written templates for various scenarios to speed up response time and ensure consistent and clear messaging.

2. **Dedicated Communication Team**: A team responsible for all communications during a crisis ensures messages are coordinated and disseminated quickly to all stakeholders.

3. **Stakeholder Mapping**: Identify all potential stakeholders (employees, media, investors, local communities, etc.) and tailor messages to meet their concerns and needs.

4. **Media Management**: Establish protocols for dealing with the media, including press releases, press conferences, and social media updates, to control the narrative and provide accurate, timely information.

5. **Regular Updates**: To maintain trust and manage anxiety, schedule updates to keep all stakeholders informed as the situation progresses, even if the status is unchanged.

### Three Cs Model: Command, Control, and Communicate

1. **Command**: Leadership takes charge of the situation, ensuring a clear chain of command to prevent confusion and mismanagement.

2. **Control**: Implement structured systems and processes to manage the crisis, including task prioritization and resource allocation.

3. **Communicate**:

   - *Internal Communication*: Keep staff and internal stakeholders well-informed, motivated, and calm.

   - *External Communication*: Manage information flow to the public, media, and other external stakeholders effectively using multiple channels such as social media, the organization's website, and traditional media to reach different audiences.

   - *Two-way Communication:* Establishing channels not just for disseminating information but also for receiving feedback, which can be crucial for adjusting strategies or correcting misinformation.

The CCP and Three C's Model emphasize the importance of preparedness, structure, and adaptability in crisis communication. These methods help organizations maintain trust and control over the situation by ensuring all stakeholders are informed and engaged throughout the crisis. By implementing these strategies, organizations can significantly enhance their ability to communicate clearly and precisely in critical moments, improving overall effectiveness and reducing the risk of errors in high-stress environments.

# Chapter 14

# EXTREME LEADERSHIP CHANGES OVER TIME

Over the centuries, many Big Visions have been accomplished—from climbing the highest mountains to reaching the bottom of the Mariana Trench, from building railroads across America to constructing the vast interstate highway network, from flying around the Earth to landing on the moon, and from discovering unknown lands to uncovering vast energy sources. All the Big Visions may seem to have been achieved, but that is far from the case. The world is still not at peace; self-serving manipulators dominate economies, and the universe remains unexplored. We have yet to learn how to travel across light-years, explore all ocean depths, or reach the Earth's core. Many species of animals and plants remain undiscovered. We still struggle to preserve racial and cultural diversity while respecting our differences, and significant religious conflicts persist. The potential for new Big Visions is endless.


### In the Past

In our early history, Big Visions were often about adventure, discovery, religious conversion, kingdom expansion, fame, and scientific breakthroughs. Today, while discovery and adventure still play a role, many Big Visions focus on financial gain—oil exploration, patents, and new internet ventures. However, the critical leadership strategies remain fundamentally the same.

Some examples of key leadership strategies that still apply today include:

1. **Visionary Thinking**: Leaders must envision the future and set ambitious goals.

2. **Decisive Action**: Quick, effective decision-making is crucial in navigating challenges.

3. **Adaptability**: Flexibility in strategies and tactics allows leaders to respond to changing circumstances.

4. **Collaboration**: Building solid teams and fostering collaboration enhances problem-solving and innovation.

5. **Resilience**: Maintaining focus and perseverance despite setbacks ensures long-term success.

6. **Ethical Leadership**: Upholding integrity and ethical standards will build trust and credibility.

By applying these timeless leadership strategies, we can continue to pursue and achieve new Big Visions in an ever-evolving world.

### The Importance of Timely Action in a Crisis

In times of crisis, taking timely and decisive action can be the key to success. Delayed responses can exacerbate problems and diminish the chances of a positive outcome. Here's why timely action is crucial and how it can be effectively implemented:

1. **Implementation Execution as a Critical Factor**

   Executing plans efficiently and effectively is paramount during a crisis. Implementation execution involves:

- *Precise and Thorough Planning*: Detailed plans must be in place and ready to be activated. These plans should outline specific steps, assign responsibilities, and establish timelines for each action. A clear roadmap ensures that everyone knows their role and what is expected of them.

- *Resource Allocation*: Ensuring that the necessary resources—such as personnel, equipment, and information—are available and can be mobilized quickly. Effective resource management can make the difference between success and failure.

- *Coordination and Communication*: Maintaining open lines of communication and ensuring coordination among all team members and stakeholders. This approach helps prevent misunderstandings and ensures everyone works towards the same goal.

- *Adaptability*: Being prepared to adjust plans as the situation evolves. Crises are often unpredictable, and the ability to pivot and adapt to new information or changing circumstances is crucial.

2. **Avoiding Reinforcement of Failure**

In a crisis, it is essential to recognize and address failures quickly without reinforcing them. Here are strategies to avoid reinforcing failure:

- *Acknowledgment of Mistakes*: Leaders must be willing to acknowledge when something is not working. This openness allows for swift correction and prevents the continuation of ineffective actions.

- *Learning from Failure*: Every failure should be seen as a learning opportunity. Analyzing what went wrong and why it happened can provide valuable insights that help improve future responses.

- *Promoting a Culture of Improvement*: Encourage a culture where continuous improvement is valued over punitive measures. This culture fosters an environment where team members feel safe reporting issues and contributing to solutions.

- *Rapid Course Correction*: It is essential to implement immediate corrective actions to address failures. This may involve changing tactics, reallocating resources, or altering plans to better align with the crisis's current realities.

Leaders can navigate crises more successfully by focusing on timely action and effective execution. Quick and decisive action, coupled with the ability to learn from and correct failures, ensures that teams can respond to challenges efficiently and effectively, improving their chances of a successful resolution.

As we have explored the evolution of extreme leadership over time, it is now crucial to examine the tangible impacts these leaders have had on their societies. Understanding the consequences of extreme leadership provides valuable insights into the lasting effects of their actions and decisions. Not only do the strategies remain the same over time, but so do many of the consequences.

# Chapter 15

# THE CONSEQUENCES OF EXTREME LEADERSHIP

The societal consequences of extreme leaders' Big Visions are often substantial, whether those visions are achieved or merely attempted. These impacts can include the ending of wars, the dominance of a group of people, the elimination of injustices, and more.

Extreme leaders operate under immense stress but rarely show it. They may have periodic doubts about choosing the right Big Vision and goals. While goals can be adjusted if a vision proves impossible and others have significantly sacrificed to achieve it, leaders must grapple with the consequences of their actions. They constantly worry about securing sufficient resources and ensuring they arrive in the nick of time. Additionally, extreme leaders must always be concerned about their followers—are they getting the necessary resources and care? Performance is another critical concern. Leaders must assess whether their followers are doing the job and decide whether to allow more time or replace underperformers. This decision carries risks, including the time required for new team members to become productive and the possibility that the change may not yield the desired results. Finally, leaders must consider their safety and exposure to risk. They cannot be cavalier about their security.

The consequences for extreme leaders often remain the same regardless of success or failure. In the end, we all die, but extreme leaders rarely die of natural causes. They may perish at the moment of victory or soon after, as did General George S. Patton, or at the moment of mission failure, like Adolf Hitler. Death may be self-inflicted, at the hands of opponents, as with Mahatma Gandhi and Martin Luther King Jr., or in retaliation from adversaries, such as Osama Bin Laden. Previous supporters or even those they saved may also turn against them, as in the case of Joan of Arc.

If death does not come at the time of success or failure, extreme leaders often find it difficult to return to a calm life. Lawrence of Arabia, Hernando Cortes, and General George S. Patton briefly found civilian life tedious, frustrating, and dull. Having lived such fast-paced and dramatic lives, the slower pace of peace can be a heavy burden. Some extreme leaders live long lives in continuous pursuit of their Big Vision, never fully achieving it to their satisfaction, and die "in the saddle," as did Mother Teresa.

Few extreme leaders live out their natural lives. Winston Churchill, Dr. Elisha Kane, Dwight D. Eisenhower, Neil Armstrong (died 2012), and Michael Collins (died 2021) are exceptions. As of July 2024, Buzz Aldrin, at age 94, still is. If extreme leaders' purposes are noble, those who survive may be lionized while still living and remembered favorably in history. Those who die for a good cause are often remembered as heroes, heroines, or saints. Conversely, those who survive with an evil purpose are remembered poorly in history and may face conviction and punishment, including execution or life imprisonment. Those who die with an evil purpose are remembered as villains for their malevolent deeds.

# CONCLUSION

Are you an extreme leader, or do you aspire to become one? If so, this roadmap provides the guidance you need. Even if extreme leadership is not your goal, you can still emulate the skills and characteristics of the extreme leaders described above to become an outstanding leader or an excellent follower.

Extreme leadership involves a unique blend of vision, decisiveness, resilience, and the ability to inspire and mobilize others under challenging circumstances. By understanding and applying these principles, you can enhance your leadership capabilities and contribute more effectively to your team's success. Whether leading a significant initiative or supporting a visionary cause, the traits and strategies outlined in this guide will serve you well in any role you pursue.

What follows are case studies that illustrate these principles in action, providing concrete examples of extreme leadership and the profound pact it can have.

# CASE STUDIES

# HERNANDO CORTES, EXPLORER

A Case Study Showing the Distinction Between
Purpose and Leadership Skills and Giving Examples
of Skills Used in Achieving Big Visions

Based on the Book

*History of the Conquest of Mexico:*

*The Life of the Conqueror Hernando Cortes*

by

William H. Prescott

With excerpts, edits, paraphrasing, and

commentary

by

Charles Patton

EXTREME LEADERSHIP

## Introduction

When reading this example of an extreme leader, it is crucial to distinguish between leaders' skills and the purposes for which they utilize them. An extreme leader can use practical leadership skills for both good and bad purposes, and history provides numerous examples across the spectrum of morality. By examining this case, we do not condone Cortes's brutal treatment of the native Aztecs. His actions were comparable to those of settlers in North America against Native Americans, often cruel, unfair, and fatal. These actions reflect the attitudes and outlooks of the time, where major political and religious powers sought to claim large portions of the New World with indiscriminate force.

The results of Cortes's actions are reprehensible, yet his methods can still be studied for applications in more appropriate contexts in today's "more civilized" world. The key takeaway is that extreme leaders have used their skills for evil and honorable purposes. Additionally, the perception of good and evil can vary depending on the leader's focus and history's judgment.

Cortes believed his actions were honorable; in the eyes of his monarchy, church, peers, and countrymen, he was a Soldier of the Cross. Only through the eyes of the vanquished and the hindsight of historians can his cruelty be entirely judged. Historian William Prescott described Cortes as "not being cruel, at least not as cruel compared with most of those who followed." Prescott also noted that Cortes was "severe in enforcing obedience to his orders for protecting the natives and their property." Although his directives were not always followed, Cortes enforced them rigorously whenever he discovered violations.

> Extreme leaders clearly understand their purpose before applying their leadership skills. Ideally, this purpose serves the greater good rather than evil.

### Cortes's Beginnings

Cortes was born in Medellin, near Seville, Spain, in 1485. His father, an infantry captain, and his mother were respected for their excellent qualities. During his youth, Cortes was feeble but grew stronger as he matured. At 14, his father sent him to Salamanca to study law. However, he showed little interest in academics. After two years, he returned home, much to his parents' disappointment. Though he learned some Latin and wrote good prose, he was otherwise idle until, at age 17, he decided to join the military for the adventurous life it promised.

He planned to sail with Don Nicolas de Ovando, Columbus's successor but fell from a high wall while trying to access a lady's apartment. This accident resulted in a severe contusion and forced him to stay in bed, causing him to miss the fleet's departure. Cortes remained home for two more years, gaining little from the experience. (Prescott, 1843, Vol I., pp. 1–467, 1–490)

### Off to Sea

In 1504, at age 19—the year Queen Isabella I of Castile, known as Isabella the Catholic, died—Cortes sailed with Alonso Quintero in a small squadron bound for the Native American islands. Queen Isabella earned her title "Isabella the Catholic" from Pope Alexander VI in recognition of her devout faith and efforts to reform the church and strengthen Christianity in Spain.

Quintero attempted to reach Hispaniola ahead of the other ships in his group but got lost in heavy gales. Eventually, a dove landed on the mast, which Cortes saw as a miraculous sign leading them to Hispaniola. Due to the delays caused by the storms, Quintero's companions arrived after him. Upon landing, Cortes visited the governor, Don Nicolas de Ovando, whom he knew from Spain. The governor persuaded Cortes to stay and granted him land in Acua. Cortes settled there for a time, often engaging in duels, and earning numerous scars from his encounters with an expert swordsman.

## Cortes Becomes a Soldier

He occasionally participated in military actions to suppress native uprisings under Ovando's lieutenant, Diego Velasquez, becoming familiar with the tactics, toil, and dangers of Native American warfare, as well as the acts of cruelty that were commonly practiced.

In 1511, when Velasquez embarked on the conquest of Cuba, Cortes abandoned his quiet life and joined the expedition. He demonstrated remarkable activity and courage throughout the invasion, earning the commander's attention. His free and cordial manners, good humor, and lively wit made him a favorite among the soldiers. However, he showed little of the discipline and seriousness that would characterize his later years. After the island's conquest, Cortes was highly favored by Velasquez, now the governor, who appointed him as his secretary.

On the island, Cortes became enamored with Catalina Xuarez from Granada, Spain. He promised to marry her but later resisted pressure from her family and the governor to fulfill this promise after becoming infatuated with one of Catalina's sisters, who reciprocated his feelings.

### Cortes Crosses the Establishment

Cortes grew distant from the governor and began associating with others dissatisfied with the distribution of lands and offices. The dissatisfied group decided to present their grievances to the authorities in Hispaniola, from whom Velasquez had received his commission. The voyage to Hispaniola, undertaken in an open boat across eighteen leagues of sea (about forty-seven miles), was hazardous, and they chose Cortes as the best man for the task.

However, the conspiracy reached the governor's ear before the envoy's departure. Velasquez seized Cortes, fettered him, and placed him in strict confinement. It was rumored that he would have hanged Cortes if not for the intervention of his friends. Cortes did not remain confined for long; he freed himself, escaped, and sought sanctuary in a church.

The governor, unwilling to violate the sanctity of the church, stationed a guard to watch for Cortes to leave. Within a few days, as Cortes stood outside the church, the guard, Juan Escudero, ambushed and captured him. Cortes later hanged Escudero in New Spain (i.e., Mexico) for his involvement in a similar plot against him. Cortes was again put in irons and placed aboard a vessel set to sail for Hispaniola.

Once more, Cortes managed to free himself, escaping into

a boat and quietly pushing off. As he neared the shore, the currents became rapid and strong. Being an excellent swimmer, he plunged into the water, struggled for his life, and successfully reached the shore, returning to the church to claim sanctuary again.

For reasons unknown, Cortes eventually relented and agreed to marry Catalina Xuarez, securing the favor of her family. Subsequently, the governor ceased his efforts to confine Cortes. A peculiar story is associated with this event: it is said that Cortes, in a proud spirit, refused to accept the reconciliation and unexpectedly presented himself, fully armed, before Velasquez in his quarters during a military excursion. Startled by Cortes's sudden appearance, Velasquez demanded an explanation. After a heated debate, they reconciled amicably. When a messenger arrived to report Cortes's escape, he found Cortes and the governor sleeping together in the governor's bed. Though the reconciliation was permanent, Cortes was not reinstated as secretary.

Cortes received an allotment of Native Americans for forced labor, known as a repartimiento, along with ample territory near St. Jago, Cuba, where he was soon made alcalde. He devoted himself to agriculture with increased zeal, introducing new cattle breeds to Cuba and operating gold mines that promised better returns than those on Hispaniola. Within a few years, he accumulated 2,000 to 3,000 castellanos, a substantial sum, albeit at a significant cost to Native American lives.

### The Gold Bug Bites – Cortes and the Governor

When another Spaniard, Grijalva, returned from Mexico with news of discoveries and the rich fruits of his trade with the indigenous population, the information spread like wildfire. Everyone saw in it the promise of more wealth than had ever been obtained. The governor resolved to explore the same areas as Grijalva but with considerable armament. He sought a suitable person to share the expense and take command. Despite considering several *hidalgos* (nobles), the governor rejected them all. However, he trusted two men, both close to Cortes. Seizing the opportunity, Cortes asked these men to recommend him as the ideal candidate to lead the expedition.

An extreme leader seizes opportunities as they arise, driven by ambition and a willingness to give everything to achieve a noble purpose. They maintain an unwavering focus on their overarching vision, dedicating themselves entirely to its realization.

Cortes is reputed to have offered his recommenders a liberal share of the expedition's proceeds. They advocated for his selection with all their eloquence. Cortes had amassed a fortune and possessed the necessary experience to lead the expedition. His popularity would also attract followers to his banner. The governor summoned Cortes and announced his intent to appoint him captain-general of the armada. Cortes had finally achieved his goal since setting foot in the New World: to claim Mexico for Spain and the pope.

Cortes saw boundless potential before him—no more struggle on a petty island. This time, it was his opportunity to fulfill his ambitions for wealth and power. From occasional rumors, he sensed Mexico was a great empire. His levity and idle merriment evaporated, replaced by intense concentration on his Big Vision. He invested all his money to outfit the armament, mortgaging everything, borrowing against the expedition's success, and using his friends' credit.

He purchased vessels, provisions, and military stores, inviting recruits with offers of financial assistance and a liberal share of the anticipated profits. Some converted their estates into money to equip themselves. The town of St. Jago buzzed with activity. Cortes procured six ships, some of significant size, and enrolled three hundred recruits in a few days. The extent of the governor's contribution is unclear, though Cortes's friends claimed it was minimal.

The expedition's objectives, according to the governor's direction, were to find men held captive in Yucatan, barter with the natives, and, most importantly, convert the indigenous population to Catholicism. Cortes was instructed to treat the natives with kindness and humanity. He was to declare his allegiance to the Royal Master and manifest it by presenting gold, pearls, and precious stones, thus securing the sovereign's favor and protection (Prescott, W. H., 1843).

Cortes was to familiarize himself with the country's natural products, the character of the races, their institutions, and their level of civilization. He was to send home detailed accounts of these findings and the articles obtained through his interactions. Finally, he was to take utmost care to omit nothing from his survey that might benefit the church or his sovereign. The governor had not received authority to colonize Mexico but only to barter with the natives.

> An extreme leader meticulously crafts a plan thar will realize their vision.

### The Governor Reconsiders Cortes's Commission

Cortes's increasing importance and loftier bearing gradually made the naturally suspicious Velasquez uneasy. He became apprehensive that his officer, once away with considerable power, might also be inclined to sever his dependence on the governor. An incident at this time heightened these suspicions.

A half-mad fellow, the governor's jester, cried out while Velasquez walked with Cortes toward the port, "Have a care, master, or we shall have to go hunting, some day or another, after this same captain of ours!" Cortes dismissed the jester's comment: "He is a saucy knave and deserves a good whipping." However, the words had already sunk deep into Velasquez's mind, and some of the governor's relatives fanned the latent embers into a blaze of jealousy. They reminded Velasquez of his past quarrel with Cortes. By misconstruing Cortes's recent actions, they convinced the governor to entrust the expedition to other hands.

The governor communicated his design to his confidential advisors, Lares, and Duero, who promptly reported it to Cortes. They advised him to expedite matters and set sail immediately to retain command. Cortes showed his usual prompt decisiveness, a trait that would often direct his destiny.

Though he had not yet assembled his full complement of men, vessels, and supplies, Cortes resolved to weigh anchor that night. He informed his officers of his plan and the reason behind it. He visited a butcher and relieved him of all his stock, leaving in payment a massive gold chain of much value

128

that he had been wearing around his neck. When the town was asleep at midnight, the men quietly boarded their vessels, and the small squadron slipped out of the bay.

The following day, the entire town was amazed. When the news reached the governor, he sprang from his bed, dressed hastily, mounted his horse, and galloped to the port, followed by his entourage. Seeing their approach, Cortes entered an armed boat and rowed within speaking distance of the shore. The governor said, "And it is thus you part from me! A courteous way of taking leave, truly." Cortes responded, "Time presses and some things should be done before they are even thought of. Has your Excellency any command?"

The mortified governor had nothing to say in return, so Cortes, politely waving his hand, returned to his vessel. On November 18th, 1518, the fleet set sail for the port of Macaca, about fifteen miles away. The governor returned to his house, realizing he had blundered in appointing Cortes and now likely made him an enemy. The clandestine departure would later be severely criticized. Despite being duly authorized by the authorities of Hispaniola, Cortes was at risk for his reputation and fortune.

Cortes felt obligated to his employer to conduct the enterprise. From Macaca, where he obtained missing stores from the royal farms as an unofficial "loan from the king," he proceeded to Trinidad, a larger town on the southernmost coast of Cuba. He landed, erected his standard in front of his quarters, and announced his mission, making liberal offers to all who would join the expedition. Volunteers came in daily, including over one hundred of Grijalva's men who had just returned from his voyage and were eager to return to Mexico.

Cortes's fame attracted cavaliers of family and distinction, including Pedro de Alvarado and his brothers, Cristobal de Olid, Alonso de Avila, Juan Velasquez de Leon (a near relative of the governor), Alonso Hernandez de Puertocarrero, and Gonzalo de Sandoval—all of whom would play important roles in the conquest. They were welcomed with lively music and joyous artillery salutes upon their arrival. Cortes acquired the military stores he needed.

He learned that a trading vessel laden with grain and other commodities for the mines was off the coast, so he ordered one of his caravels to seize it and bring it into the port. He paid the master in promissory notes for cargo and ship and persuaded him, a wealthy man named Seldeno, to join the expedition. He then dispatched Diego de Ordaz to seize another ship similarly and meet him off Cape St. Antonio, the island's westernmost point. This maneuver temporarily distanced Ordaz, who was linked to the governor and thus an inconvenient spy on Cortes's actions.

While Cortes was busy with these preparations, the commander of Trinidad received letters from Velasquez ordering him to seize and detain Cortes. The commander informed Cortes's principal officers, who counseled him against attempting to seize Cortes, fearing it could cause a commotion among the soldiers that might leave the town in ashes. Cortes ignored the letters.

He ordered Alvarado, with a small body of men, to march cross-country to Havana while he would sail around the island's westernmost point and meet him there with the squadron. Upon arrival, he prepared his ships and men better, equipping them with thickly quilted jackets that served as body armor. He then organized the men into eleven companies, each under the command of an experienced officer, including some who were friends or kin of Velasquez, treating them no differently than the others.

His principal standard was black velvet embroidered with gold and emblazoned with a red cross amidst flames of blue and white, bearing the Latin motto, "Friends, let us follow the cross; under this sign, if we have faith, we shall conquer." He increased the number of domestics and officers in his household, placing himself on the same footing as a man of high station—a state he would maintain for the rest of his life.

At age 33 or 34, Cortes was of average size, with a pale complexion and large dark eyes that gave him a grave expression at odds with his cheerful temperament. He was slender with a deep chest, broad shoulders, and a muscular frame, exhibiting a combination of vigor and agility that qualified him in fencing, horsemanship, and other chivalrous exercises. He was careful about his diet, drank little, and was indifferent to toil and privation. His dress was rich but not flashy or striking. His frank and soldier-like manners concealed a calm and calculating spirit.

Even in his gayest humor, a settled air of resolution made those who approached him feel compelled to obey, infusing something like awe into the attachment of his most devoted followers. This combination of love tempered by authority inspired devotion among the rough and turbulent spirits with whom his lot was cast. His character evolved with his changing circumstances, calling forth new qualities from within him.

> Extreme leaders maintain an unwavering air of resolution, making prompt decisions at critical moments. They refuse to let obstacles impede their progress and are willing to adapt their character to meet changing circumstances.

### Cortes Sneaks out of Town

Before Cortes's preparations were complete, the commander of Havana, Don Pedro Barba, received dispatches from Velasquez, ordering him to apprehend Cortes and prevent the departure of his vessels. Cortes also received a letter requesting him to postpone his voyage until the governor could communicate with him in person. Despite this, Cortes had the goodwill of Barba, who lacked the power to restrain him, especially since Cortes's officers and men were prepared to lay down their lives for him.

Cortes wrote to Velasquez, imploring him to trust in his devotion to the governor's interests, and concluded with the firm assurance that he and the whole fleet, God willing, would set sail the following morning. On February 10th, 1519, the squadron, consisting of eleven ships (one of one hundred tons, three of 70 to 80 tons, and the remainder being caravels and open brigantines), got underway and met the others at Cape St. Antonio.

The entire fleet was placed under the chief pilot, Anton de Alaminos, a veteran navigator who had served Columbus, Cordova, and Grijalva in his former expedition to the Yucatan. After landing on the cape and mustering his forces, Cortes found he had 110 mariners, 553 soldiers (including thirty-two crossbowmen and thirteen arquebusiers), two hundred Native Americans from the island, and a few Native American women for menial chores. He was also provided ten heavy guns, four lighter guns called falconets, and a good supply of ammunition. Though this may sound like a formidable force, the native populations Cortes would face numbered in the hundreds of thousands.

Cortes also had sixteen horses, which were difficult to transport in the small ships. Still, he had correctly estimated the importance of cavalry for their effectiveness in the field and their ability to strike terror into the natives. Cortes embarked on the conquest with a stout heart, though the force would have seemed paltry to him had he foreseen half of the difficulties ahead.

Cortes addressed his men with a short but impassioned speech, telling them they were embarking on a noble enterprise that would make their names famous for ages. He emphasized that they were heading to lands only a handful of Spaniards had ever seen:

"I hold out to you a glorious prize, but it is to be won by incessant toil. Great things are achieved only by great exertion, and glory is never the reward of sloth. If I have labored hard and staked my all on this undertaking, it is for the love of that renown, which is the noblest recompense of man. But if any among you covets riches more, be but true to me, as I will be true to you and to the occasion, and I will make you masters of such wealth as our countrymen have never dreamed of! You are few in number, but strong in resolution; and if this does not falter, doubt not that the Almighty, who has never deserted the Spaniard in his conquest of the Infidel, will shield you, though encompassed by a cloud of enemies; for your cause is a just cause, and you are to fight under the banner of the Cross. Go forward then, with alacrity and confidence, and carry to a glorious issue the work so auspiciously begun." (Prescott, W. H., 1843).

His appeal struck chords of ambition, avarice, and religious zeal, sending a thrill through the hearts of his martial audience. They received it with acclaim and eagerly pressed forward under a leader who promised not just to battle but triumph. Cortes found that his followers broadly shared his enthusiasm. Mass was celebrated, and the fleet was protected by Saint Peter, Cortes's patron saint. They weighed anchor and departed on February 18th, 1519.

During a storm, the ships separated, and Cortes was the last to arrive at the island of Cozumel. Upon arrival, he discovered that Pedro de Alvarado had entered the temples, rifled them of their few ornaments, and terrified the natives with his violent conduct, causing them to flee into the island's interior. Incensed by Alvarado's actions, which contradicted his policy, Cortes reprimanded his officer in the army's presence. He then commanded that two Native Americans captured by Alvarado be brought before him.

With the help of an interpreter, Melchorejo, whom Grijalva had brought to Cuba, Cortes explained to them that his visit was peaceful. He gave them presents and invited their countrymen to return to their homes without fear. Soon, the natives returned, and trade began, exchanging trinkets and cutlery for gold ornaments, with each party believing they had outwitted the other.

While Cortes sent Diego de Ordaz with two brigantines to search the Yucatan coast for previously captured Christians, he investigated the interior of the island. He discovered a more advanced civilization than he had expected.

> The essence of Cortes and other extreme leaders is captured in the belief that "Great things are achieved only by great exertions, and glory is never the reward of sloth." He inspired his men by convincing them they were embarking on a noble enterprise, appealing to their ambition, avarice, and religious zeal.

### Cortes Lands in Mexico

In 1519, Cortes landed on the coast of Mexico and established a colony, pressing his exploration toward what is now Mexico City. A story arose, commonly told, that he burned his ships to focus his men on the mission ahead and eliminate thoughts of returning to Cuba. Recently, he had hanged two men and mutilated another for conspiring to escape with one of the ships; notably, one of those hanged was the guard who had apprehended him outside the church. This action should have been sufficient to deter mutiny, but removing the ships from the equation further eased his mind.

In truth, Cortes burned all but one small ship, but first, he salvaged their cordage, sails, iron, and all other movable items. He burned the ships not solely to focus his men, although that was one consequence, but because heavy gales had grievously racked them and were worm-eaten, rendering them unseaworthy. Some were already struggling to stay afloat. Worm infestations in ships' wood were a common plague in those days, limiting their lifespan (Prescott, 1843, vol. I, pp. 254–255). Cortes later built new ships when it came time to return to Cuba.

There is much more to Cortes's story, but I will end here, having illustrated his leadership style. He was not overly scrupulous but was effective in executing his plans, though his fame was darkened by actions that even his boldest apologists would find hard to vindicate. (Prescott, 1843, vol. II, pp. 372–373).

> As an extreme leader, Cortes crafted grand visions for those he served—his church and crown. He swept away all obstacles in his path, made fast, bold decisions, when necessary, planned meticulously when time allowed, and accomplished his goals with remarkable speed and efficiency.

# CHRISTOPHER COLUMBUS

A Case Study of what a Big Vision and Persistence Can Achieve

Based on the Book

*A History of the Life and Voyages of Christopher Columbus*

by

Washington Irving

With excerpts, edits, paraphrasing, and commentary

by

Charles Patton

EXTREME LEADERSHIP

## Introduction

The following is excerpted and paraphrased from a marvelous account of Columbus by Washington Irving, published in 1829. Irving's work is based on his review of the original ships' logs from Columbus's voyages, which led to the discovery of the new continent that would become North, Central, and South America, and the Caribbean, where Columbus first landed. (Irving, 1829).

## His Vision

In the early 1470s, Christopher Columbus supported his family by making maps and charts. For a time, he lived on the newly discovered island of Porto Santo, where he heard fabulous tales of islands lying west of the Canary Islands. These tales ignited his imagination and led to his grand vision. Columbus formulated three core ideas supporting his radical belief that the eastern regions of Asia could be reached by sailing west. These were the bases for his core arguments:

The state of the world's understanding at the time was limited. By comparing Ptolemy's globe to an earlier map by Marinus of Tyre, Columbus divided the equator into 24 hours of 15 degrees each, totaling 360 degrees. He noted that the ancients knew that 15 hours had passed since the Canary Islands to the city of Thinae in Asia (the eastern limits of the known world at that time). With the Portuguese having advanced the western frontier by discovering the Azores and Cabo Verde Islands, Columbus estimated an additional hour. By his calculations, eight more hours, or one-third of the world, remained undiscovered.

1. **The authority of learned writers**: Columbus cited the opinions of renowned scholars such as Aristotle, Seneca, and Pliny, who suggested that one might pass from Cadiz to the Indies in just a few days. He also referenced Strabo, who observed that the ocean surrounded the earth. Furthermore, he drew on the narratives of Marco Polo and John Mandeville, who traveled east in the thirteenth and fourteenth centuries, describing Asia as extending far to the east. Additionally, Columbus referenced a letter from Paulo Toscanelli, a Florentine doctor, to Fernando Marinex sent to Columbus in 1474. Toscanelli maintained that Columbus could reach India by sailing west for 4,000 miles from Lisbon to Mangi, near Cathay (the northern coast of China).

2. **Reports of respected navigators**: Columbus gathered evidence from respected navigators suggesting the existence of land to the west, inferred from items that had floated to the shores of the known world. For instance, Martin Vincent, a pilot in the service of the king of Portugal, reported finding a piece of carved wood drifting from the west, 450 leagues from Cape St. Vincent. This wood had not been worked with an iron instrument. Similarly, Pedro Correa, Columbus's brother-in-law, cited seeing a comparable piece of wood on the island of Porto Santo, which had also drifted from the west. Columbus also heard from the King of Portugal about massive reeds that had floated to some of the islands, which Columbus believed matched Ptolemy's descriptions of reeds from India. Inhabitants of the Azores reported trunks of large pine trees, not native to the area, and the bodies of two men with features unlike those of any known race on Earth.

From these three core arguments, Columbus concluded that undiscovered land lay in the western part of the ocean and was attainable, fertile, and inhabited. The success of his undertaking depended on two errors of the most learned and profound philosophers – that Asia extended far to the east and the earth was smaller than commonly thought.

Once Columbus had his Big Vision, it influenced his entire character and conduct, and he never spoke of it with any doubt. He also was moved by the opportunity to bring the true faith to pagan lands.

> To qualify as extreme, a leader must establish a bold vision—often so grand that only he or she believes in it. An extreme leader must be fully committed and focused on achieving their Big Vision, undeterred by a lack of resources, support, or insurmountable obstacles.

### His First Attempt at Garnering Support

Years passed before Columbus could pursue his idea due to his poverty and the belief that he needed the approval of King John II of Portugal, who was engaged in the exploration of Africa at the time.

Even though the compass was in common use, mariners rarely ventured out of sight of land. Around this time, in the early 1470s, King John II called upon the most able astronomers and cosmographers of his kingdom to devise means to enhance navigation. They developed the astrolabe, the forerunner of today's quadrant, allowing seamen to ascertain their distance from the equator by measuring the sun's altitude. This innovation freed ships from the coast and enabled them to retrace their course on the return trip.

King John II was interested in finding a shorter route to India by sailing west and met with Columbus, who described the immense riches of Cipango, the first island he expected to reach. The king listened but was reluctant to back the new scheme due to the cost and the ongoing African exploration. However, Columbus's arguments were solid, so the king consented. Columbus then demanded high and honorable titles and rewards worthy of his deeds and merits. The king, perceiving Columbus as vainglorious and given to fantastic fancies, referred the matter to Rodrigo Joseph and his confessor Diego Ortiz de Calzadilla, Bishop of Ceuta. They deemed the project extravagant and visionary, considering it irresponsible and impractical.

The king then referred the matter to his council, which condemned Columbus's proposal. They argued that the nation was already engaged in a war against the Moors of Barbary and battling pestilence, and another venture would drain resources and divide the nation's power. King John II heeded his counselors, who proposed keeping Columbus in suspense while they dispatched a ship in his proposed direction, using maps Columbus had submitted for review. The ship encountered storms and returned, reinforcing the counselors' ridicule of Columbus's project as extravagant and irrational. Columbus was indignant when he learned what they had done and that they had used his maps. After his wife's death, he departed secretly from Lisbon in 1484, taking his son Diego to Genoa.

EXTREME LEADERSHIP

Columbus left Portugal owing money and could not return. After about a year, he presented his proposal to the Republic of Genoa, but they declined due to the impact of war. They encouraged him to present his proposal to Venice, even though Venice had been at war with Genoa. Meanwhile, Columbus sent his brother to present his proposal to King Henry VII of England, but his brother became distracted by the discovery of the Cape of Good Hope.

Destitute, Columbus arrived in Spain with his son Diego on his way to Huelva to seek his brother-in-law. He stopped at an ancient convent of Franciscan friars dedicated to Santa Maria de Rabida to ask for a bit of bread and water for his son. The prior, Friar Juan Perez de Marchena, noticed their demeanor and accent and soon learned their story.

The Prior, knowledgeable about geographical and nautical science from his proximity to Spain's eminent navigators in Palos and his travels to recently discovered islands and countries on the African coast, was struck by Columbus's views. He invited Columbus and his son to stay at the Priory and sent for Garcia Fernandez, the physician of Palos, who also became interested. Veteran Palos navigators provided hints that corroborated Columbus's theory. One, convinced of the enterprise's importance, Pedro de Velasco offered to introduce Columbus to the court, being on intimate terms with Fernando de Talavera, Prior of the monastery of Prado and confessor to the queen. He advised Columbus to present his proposal to the Spanish sovereigns. The prior gave Columbus a letter of recommendation to Talavera and agreed to take charge of Columbus's son and educate him at the convent while Columbus pursued his audience.

An extreme leader needs resources, often obtained only by convincing others to believe in the Big Vision. This acquisition is achieved by crafting a compelling presentation and seeking out potential audiences who will listen to the case for the Big Vision. These audiences may not have the necessary resources themselves but can be valuable allies if they know others who do.

### In Search of Support and Resources

In the spring of 1486, with fresh hopes, Columbus left the Priory and set off with his letters of recommendation. It was a brilliant time for the Spanish monarchy. Ferdinand and Isabella's marriage had united Aragon and Castile, consolidating the Christian powers. The Spanish army was laying siege to the last Moorish stronghold in the mountains around Granada.

While Ferdinand and Isabella were independent monarchs, their interests were aligned. Ferdinand was hardy, even majestic, with a clear and comprehensive genius, a grave and certain judgment, devout in his religion, and unrelenting in business. He was a keen observer and judge of men, unparalleled in the science of politics. Some saw his policies as cold, selfish, and artful, and others as wise and prudent. His three goals were to conquer the Moors, expel the Jews, and establish the Inquisition in his dominions. He achieved all three and was rewarded by Pope Innocent VIII with the appellation of Most Catholic Majesty.

On the other hand, Isabella was well-formed with great dignity and gracefulness, combining gravity and sweetness. She had blue eyes, auburn hair, a benign expression, firmness of purpose, and earnestness of spirit. She exceeded Ferdinand in genius, attractiveness, personal dignity, and grandeur of soul. She took part in his war counsels, engaged in his enterprises, and sometimes surpassed him in firmness and fearlessness while being inspired by a truer idea of glory. She focused on reforming laws, healing the ills from protracted wars, and governing the people. While influenced by her religious advisers, she remained hostile to measures advancing religion at the expense of humanity. She opposed the expulsion of the Jews and the establishment of the Inquisition, but her confessors slowly vanquished her repugnance. She advocated clemency for the Moors while being the soul of the war against Granada, and she assembled the ablest men in literature and science as her counselors, promoting art and knowledge.

Columbus arrived in Cordoba but was disappointed in his hopes of immediate sponsorship, finding it impossible to get a hearing. Prior of Prado, Fernando de Talavera viewed Columbus's plan as extravagant and impossible. As a stranger dressed in humble garb offering magnificent speculations, Talavera did not believe Columbus. At the same time, the war on Granada was at its peak, and the court was in the field, shifting from place to place, with little time to consider foreign exploration.

Columbus stayed in Cordoba until the fall, supporting himself by designing maps and charts. He slowly gained converts and friends, remaining enthusiastic and maintaining his dignity. Alonzo de Quintanilla, comptroller of the finances of Castile, became an influential advocate of Columbus's theory. Columbus also became acquainted with Antonio Geraldini, the pope's nuncio, and his brother, Alexander Geraldini, preceptor to the younger children of the king and queen.

With the aid of these two friends, Columbus was introduced to the celebrated Pedro Gonzalez de Mendoza, archbishop of Toledo and grand cardinal of Spain, the most important person at court. Always at the side of the king and queen, Gonzalez was respected for his clear understanding, eloquence, judiciousness, and excellent quickness and capacity in business. He was an elegant scholar but unskilled in cosmology. Initially, Gonzalez found Columbus's theory heretical, incompatible with the sacred scriptures' description of the earth. However, further explanation convinced him there was nothing irreligious in attempting to expand human knowledge and ascertain the works of creation. Gonzalez saw the grandeur of the conception and yielded to Columbus's arguments in a noble, earnest manner.

The grand cardinal arranged an audience with the sovereigns. Columbus appeared before them modestly but with the feeling that he was an instrument in heaven's hand to accomplish its grand designs. Ferdinand perceived the scientific and practical foundation of Columbus's soaring imagination and magnificent speculations. Excited but cautious, Ferdinand decided to seek the opinion of the most learned men in the kingdom and to be guided by their decisions.

The king authorized Fernando de Talavera to assemble the most learned astronomers and cosmographers to hold a conference with Columbus, consult together, and then make their report. Columbus was lodged and entertained with great hospitality during this examination at the Dominican convent of St. Stephen in Salamanca. The examiners were professors of astronomy, geography, mathematics, and other branches of science, along with various church dignitaries and learned friars. Columbus presented and defended his conclusions before them.

> An Extreme Leader possesses unwavering confidence and conviction in the Big Vision and diligently prepares for significant opportunities. Often, a leader gets only one "bite at the apple," meaning the right audience may only entertain the Big Vision once. Being prepared and in the right place at the right time is often seen as luck, but in reality, it is the result of meticulous planning and thorough preparation.

### The Arguments Presented Against His Vision

The challenges Columbus faced during the examination of his theory were formidable and presented by influential men. The most powerful examiners were against him. Some viewed Columbus as an impractical adventurer, while others were impatient with his ideas. Nevertheless, Columbus pleaded his case with natural eloquence. Initially, only the friars of St. Stephen paid attention to him.

In contrast, others remained entrenched in the belief that an ordinary man could not propose such a monumental discovery after thousands of years of the status quo. Columbus was a victim of monastic bigotry and the imperfect state of science at that time. He was assailed with citations from the Book of Genesis, the Psalms of David, the Prophets, the Epistles, and the Gospels, as well as expositions from Saints Chrysostom, Augustine, Jerome, Gregory, Basil, Ambrose, and Lactantius.

They attacked Columbus's citation of Pliny, who described the southern hemisphere as the exact opposite of the northern one. They argued this would mean that those nations were not descended from Adam, which would discredit the Bible, as it states all men are descended from one common parent. They pointed to the Psalms, which describe the heavens as being extended like a hide, and to St. Paul, who compared the heavens to the covering of a tent, inferring that the earth was flat. Columbus, a devoutly religious man, was in danger of being convicted not merely of error but of heresy.

Others more versed in science admitted the rounded form of the earth and the possibility of an opposite, inhabitable hemisphere. However, they argued that it would be impossible to reach due to the insupportable heat of the torrid zone, and even if the heat could be endured, the voyage would require at least three years because of the earth's circumference. They contended that anyone undertaking the journey would perish from hunger and thirst, as carrying provisions for so long would be impossible. Based on Epicurus's authority, Columbus was told that even if the earth was spherical, only the northern hemisphere was habitable and covered by the heavens. In contrast, the opposite half was chaos, a gulf, or a mere waste of water.

One of the more absurd objections raised was that even if a ship succeeded in reaching the extremity of India by this route, it would never be able to return. They argued that the globe's rotundity would present a kind of mountain impossible to sail up and over, even with the most favorable wind.

An Extreme Leader will face strong resistance from naysayers, disbelievers, and saboteurs. The leader must anticipate these challenges as much as possible and remain steadfast in the face of bluster, innuendo, and attacks.

### Columbus's Counterarguments

Columbus argued in response to the spiritual objections that the inspired writers were not speaking technically as cosmographers but figuratively in language accessible to all levels of comprehension. He treated the commentaries of the church fathers with deference, viewing them as pious homilies rather than philosophical propositions that needed to be either accepted or refuted. He met the objections drawn from ancient philosophers boldly and ably on equal terms, demonstrating his deep knowledge of cosmography. Columbus showed that the most illustrious sages believed both hemispheres to be habitable, although they imagined the Torrid Zone precluded travel. He countered this by citing his voyage to St. George la Mina in Guinea, under the equatorial line, where he found the region not only traversable but also abounding in population, fruits, and pasture. Despite being daunted by the task and the noble nature of his listeners, Columbus's religious conviction gave him confidence. Others noted his commanding presence, elevated demeanor, an air of authority, bright eyes, and persuasive intonation.

Diego de Deza, a learned Dominican friar and professor of theology at St. Stephen, saw the wisdom in Columbus's arguments and calmed the blind zeal of his more bigoted brethren, securing at least an impartial hearing for Columbus.

The most difficult issue was reconciling Columbus's plan with Ptolemy's cosmography, which placed the earth at the center of the universe—a model in which all scholars had implicit faith, despite Copernicus devising a new heliocentric model at that time. Prejudice persisted, and even the more liberal and intelligent felt little interest in the discussions. Those who listened often regarded the plan as a delightful but impractical vision.

Fernando de Talavera, entrusted with the matter, had little esteem for it and was too preoccupied with public concerns to press it to a conclusion. Thus, the inquiry experienced continual procrastination and neglect. The board's consultations at Salamanca were interrupted by the court's departure to Cordova early in the spring of 1487 for the campaign against the rugged area around Malaga. Fernando de Talavera, now bishop of Avila, traveled with the queen as her confessor, leaving Columbus in suspense.

Consideration of Columbus's proposals ebbed and flowed with the king's war. Whenever the court had leisure, they revisited his case, but the urgency of war soon swept the matter away again. In the meantime, Columbus was ridiculed as a dreamer, adventurer, and madman. Children pointed to their heads as he walked by, implying he was crazy.

However, Columbus made an impression on several learned men, including his friend Friar Diego Deza, who tutored Prince Juan. Despite the Junto of Salamanca's skepticism, favor for Columbus grew at court. Fernando de Talavera was eventually commanded to inform Columbus, then in Cordova, that the great cares and expenses of the war made it impossible to engage in any new enterprises. They promised to consider his proposal once the war concluded.

Unwilling to receive a reply from someone who had always been unfriendly, Columbus went to the court in

Seville to learn his fate from the monarchs themselves. Their response was the same—declining the enterprise for the present but holding out hope for the future. Columbus saw their answer as merely an evasive way to dismiss him. He believed the sovereigns were swayed by ignorant and bigoted objections, gave up hope of help from the crown, and left Seville in disappointment and indignation.

Despite his frustration, Columbus did not sever his ties with Spain. He had formed a tender connection with a lady from Cordova named Beatrix Enriquez, of a noble family. Though their relationship was not formalized through marriage, she became the mother of his second son, Fernando, who would later become Columbus's historian and whom Columbus treated on equal terms with his legitimate son, Diego.

An Extreme Leader must be prepared to handle all objections with solid counterarguments. The best preparation is to practice in front of trusted, intelligent friends who can simulate the most extreme attacks. This approach is comparable to how a CEO prepares for an annual shareholder meeting—anticipating all possible questions and preparing answers for each.

### Columbus Tries Another Angle

Despairing of the court, Columbus turned to wealthy Spanish nobles, including the dukes of Medina Sidonia and Medina Celi. Both dukes had estates resembling principalities, complete with ports and shipping. They served the monarchs more as princes than vassals, bringing armies of their retainers to the field, led by their captains or themselves. They supported the war effort with their armadas and treasures. Sidonia, for instance, sent a large force of cavaliers, 20,000 Doblas of gold, and one hundred vessels, some armed and others laden with provisions.

Columbus had many interviews and conversations, first with Medina Sidonia, but came to no conclusions. The magnificent potential Columbus initially tempted the duke described, but its splendor implied exaggeration. He finally rejected the venture as the dream of an Italian visionary.

Columbus next turned to the Duke of Medina Celi, who initially showed great interest. At one point, the duke was on the verge of dispatching Columbus on the contemplated voyage with three or four caravels. However, fearing the Crown's unfavorable reaction, he abruptly abandoned the offer. The venture was deemed too grand for a mere subject and fit only for a sovereign power. Nonetheless, he encouraged Columbus to reapply to the monarchs and offered to use his influence with the queen.

In the meantime, Columbus received an encouraging letter from the king of France and quickly prepared to go to Paris. Before leaving, he visited La Rabida to see his eldest son, Diego, still under the care of his zealous friend, Friar Juan Perez. After seven years of soliciting at the court, Columbus arrived at the friar's doorstep, moved by his poverty and disappointment. Perturbed by the potential loss of such an essential enterprise to another country, the friar consulted his learned friend Garcia Fernandez and Martin Alonzo Pinzon, head of the wealthy and distinguished family of Navigators of Palos.

Pinzon agreed to support Columbus's plan with his purse and presence, offering to bear Columbus's expenses in a renewed application to the court. Knowing the queen well from his time as her confessor, Friar Perez believed she was accessible.

Columbus, reluctant to leave Spain, which he felt was his home, also hesitated to face the same vexations and disappointments in another court as he had in Spain and Portugal. The small council chose Sebastian Rodriguez, a pilot of Lepi and one of the shrewdest and most influential people in that maritime neighborhood, to act as an ambassador.

> An Extreme Leader must be resilient and persistent. While it is easy to become discouraged, staying firm in resolve is a crucial trait that separates extreme leaders from ordinary leaders.

## Columbus Tries Again

The queen was in Santa Fe, the military city built in the Vega before Granada. Their ambassador, Rodriguez, swiftly gained access and delivered Columbus's letter from the friar to the queen. Isabella had already been favorably inclined toward Columbus's proposition, having also received a recommendation from the Duke of Medina Celi. Fourteen days later, she wrote back, asking Juan Perez to come to the court, giving Columbus renewed hope.

Before midnight on the day the letter was received, the friar saddled his mule and set off privately for the court. He rode through the recently conquered Moorish territories to Santa Fe. The queen admitted him, and he pleaded

Columbus's case with characteristic enthusiasm, highlighting Columbus's professional knowledge, experience, honorable motives, and capacity to fulfill the undertaking. He emphasized the solid principles on which the undertaking was founded, the advantages of its success, and the glory it would bring to the Spanish Crown.

Isabella, never having heard such honest zeal and impressive eloquence, was moved. More easily influenced than the king, she was further swayed by the support of her favorite, the Marchioness of Moya. The queen requested Columbus to be sent to her and, aware of his humble situation, ordered 20,000 maravedis in florins ($72, or $216 in 1829 dollars) to be sent to Columbus. This sum was sufficient for his traveling expenses, a mule, and decent clothes to appear respectable at court.

The friar delivered the money and a letter to the physician Garcia Fernandez, who then handed them to Columbus. Columbus purchased a mule and clothes and set out immediately for the camp before Granada.

Upon arrival and under the care of his steadfast friend Alonzo de Quintanilla, Columbus witnessed the memorable surrender of Granada to the Spanish forces. He watched as Boabdil, the last of the Moorish kings, rode out from his seat of power while the king and queen advanced in a proud and solemn procession to receive the keys. After eight hundred years of struggle, the Moorish crescent flag was cast down, and the Spanish standard flew from the highest tower of the Alhambra.

The triumph was a victory of arms, Christianity, and the Spanish monarchy. It marked Spain's salvation and resurgence. The court was thronged with the most illustrious nobles, dignitaries, and pious figures, accompanied by bards, minstrels, and the retinue of a romantic and picturesque age.

Glittering arms, rustling robes, and music and festivity abounded.

Columbus was described as "A man obscure and but little known followed at this time the court. Confounded in the crowd of importune applicants, feeding his imagination in the corners of antechambers with the pompous project of discovering a world; melancholy and dejected in the midst of the general rejoicing, he beheld with indifference, and with contempt, the conclusion of a conquest which swelled all bosoms with jubilee, and seemed to have reached the utmost bounds of desire. That man was Christopher Columbus." (Irving, W. ,1829)

The moment arrived when the monarchs were ready to attend to Columbus's proposals. True to their word, they appointed trusted individuals to negotiate with him, including Fernando de Talavera, now Bishop of Granada, following the conquest. However, at the outset of their negotiations, unexpected difficulties arose. Columbus, fully committed to the enterprise, insisted on nothing less than princely conditions.

> An Extreme Leader will inevitably face ups and downs. By accepting these fluctuations as an essential part of the journey and maintaining persistence, the leader keeps the door open for success.

### Columbus Negotiates

His principal stipulation was that he be invested with the titles and privileges of Admiral and Viceroy over the lands he might discover and receive one-tenth of all gains from trade or conquest. The courtiers who heard his proposals were indignant, their pride shocked to see a man they considered a needy adventurer aspiring to ranks and dignities greater than their own. One observer noted that Columbus had nothing to lose in case of failure.

In response, Columbus offered to furnish one-eighth of the cost on the condition of enjoying one-eighth of the profits. His terms were deemed inadmissible. Fernando de Talavera, who had always viewed Columbus as a dreaming speculator or a needy applicant for bread, was astonished and indignant. To see a man who had been indigent and threadbare for years assume such a lofty tone and claim an office approaching the dignity of the throne was too much for the prelate.

He argued to Isabella that lavishing such distinguished honors upon a nameless stranger would degrade the Crown. Even in the case of success, such terms would be excessive; in the case of failure, they would reflect poorly on the Crown.

Isabella, always attentive to her advisers, thought the proposed advantages might be too expensive. She offered Columbus more moderate terms, presenting them as highly honorable and advantageous. However, Columbus would not cede a single point in his demands, and the negotiation was broken off. One must admire Columbus's unwavering constancy of purpose and loftiness of spirit.

> An Extreme Leader will not shy away from the core purposes behind the Big Vision and will not undervalue its potential success.

## After 18 Years, Columbus Gives Up

More than 18 years had passed since Columbus's first correspondence with Paolo Toscanelli of Florence. During that period, he endured poverty, neglect, ridicule, and countless hours seeking support from various courts. Yet nothing could shake his perseverance or make him compromise on terms he deemed beneath the dignity of his enterprise. He ignored his current obscurity and poverty as his fervent imagination grasped the magnitude of his contemplated discoveries and the potential for negotiating an empire.

Indignant at the repeated disappointments in Spain, Columbus resolved to abandon his efforts there rather than compromise his demands. In February 1492, he bid farewell to his friends, mounted his mule, and left Santa Fe for Cordova, intending to depart immediately for France. Realizing his determination, a few zealous friends made one last effort to prevent Columbus and his plan from being lost.

Luis de Santangel, the receiver of the ecclesiastical revenues in Aragon, accompanied by Alonzo de Quintanilla, secured an immediate audience with the queen. The urgency of the moment fueled Santangel's courage and eloquence. He expressed astonishment that a queen who had undertaken so many tremendous and perilous enterprises would hesitate over one where the potential loss was minimal, but the gain could be incalculable.

He reminded her of God's glory, the church's exaltation, and the extension of her power and dominion that the enterprise promised. He warned of the regret and triumph for her enemies and sorrow for her friends if the enterprise succeeded for another power. He highlighted the fame and dominion other princes had gained through their discoveries and urged her not to be swayed by learned men dismissing the project as a visionary's dream.

Santangel supported Columbus's judgment and the soundness and practicality of his plan, arguing that even failure would not disgrace the Crown. The endeavor was worth the trouble and expense to resolve even a doubt on such a significant matter, as it was the duty of enlightened and magnanimous princes to investigate and explore the universe's wonders and secrets. He noted Columbus's offer to bear one-eighth of the expense and stated that the entire enterprise required only two vessels and 300,000 crowns.

The Marchioness of Moya also exerted her influence, and Isabella's generous spirit was ignited. She resolved to undertake the enterprise but hesitated momentarily, as the king was indifferent, and the war strained the royal finances. How could she draw on an exhausted treasury for an endeavor the king opposed?

As Santangel trembled, the queen, with enthusiasm worthy of herself and the cause, declared, "I undertake the enterprise for my own crown of Castile and will pledge my jewels to raise the necessary funds." This moment defined Isabella forever as the patroness of the discovery of the New World. To secure her decision, Santangelo assured Her Majesty she would not need to pledge her jewels, as he was ready to advance the necessary funds.

She gladly accepted his offer. The funds, totaling 17,000 florins ($ 3.7 million in 2024 dollars), came from the Aragon coffers. Some years later, the king used part of the gold brought by Columbus from the New World to gild the vaults and ceilings of the grand palace of Zaragoza in Aragon.

The queen dispatched a messenger on horseback to call Columbus back. He was overtaken two leagues from Granada at the bridge of Pinos, a mountain pass famous for bloody encounters between Christians and infidels during the Moorish wars. Initially hesitant to subject himself to further delays and equivocations of the court, Columbus was informed of the queen's ardor and positive response. Thus, he returned to Santa Fe.

> An Extreme Leader must recognize that there may come a time when giving up seems like the only option. It is precisely at this moment that an Extreme Leader must persevere and make another attempt at achieving their objectives. This never-say-die attitude is what sets extreme leaders apart. Whether Columbus exemplified this or not is uncertain. There is a principle in sales called the "takeaway." When a customer is reluctant to buy, suddenly withdrawing the offer can often make them drop their resistance and want to buy. Perhaps Columbus intuitively understood this strategy.

### Columbus Gets His Way

Upon arriving at Santa Fe, Columbus was granted an immediate audience with Queen Isabella. Her kindness and support dispelled all doubts and difficulties, atoning for past neglect. King Ferdinand's concurrence was readily attained, thanks to the mediation of his grand-chamberlain and favorite, Juan Cabrero, but primarily due to his deference to the queen's zeal. While Isabella was the driving force behind this grand enterprise, Ferdinand remained cold and calculating.

Columbus presented his venture as an excellent opportunity to propagate the Christian faith, expecting to reach the vast and magnificent empire of the Grand Khan in

Asia, as described by Marco Polo. He reminded the monarchs of the Grand Khan's previous inclination to embrace Christianity and the missions sent by popes and pious sovereigns to instruct him and his subjects in Catholic doctrines. Columbus envisioned extending the light of revelation to the earth's remotest ends.

Ferdinand, however, was not swayed by religious reasoning. He saw extending the church's sway as a means to expand his dominions, as evidenced by the conquest of Granada. According to the doctrines of the day, any nation refusing to acknowledge Christianity was fair game for a Christian invader. Ferdinand was more motivated by Columbus's accounts of the wealth of Mangi, Cathay, and other provinces under the Grand Khan than by the prospect of converting him and his subjects.

On the other hand, Isabella was filled with pious zeal and motivated by the idea of accomplishing a great work of salvation. When Columbus later departed, he was given letters for the Grand Khan of Tartary.

Columbus's enthusiasm extended further. With open communication with the Crown and the anticipation of boundless wealth from his discoveries, he suggested dedicating the wealth brought back to rescue the Holy Sepulcher of Jerusalem from the infidels. The sovereigns, amused by his imagination, expressed approval, and assured him they would support that holy undertaking, even if funds were not immediately available.

This mission to recover the Holy Sepulcher was a cherished design of Columbus, one he carried throughout his life and solemnly provided for in his will. He felt chosen by heaven as the agent to accomplish this.

Juan de Coloma, the royal secretary, drew up articles of agreement with the following terms:

160

Columbus would hold the office of admiral for life and his heirs and successors forever in all the lands and continents he might discover or acquire in the ocean, with similar honors and prerogatives to those enjoyed by the high admiral of Castile in his district.

1.  Columbus would be Viceroy and Governor-General over all the said lands and continents, with the privilege of nominating three candidates for the government of each island or province, one of whom would be selected by the sovereigns.

2.  After deducting costs, Columbus would be entitled to one-tenth of all pearls, precious stones, gold, silver, spices, and other articles and merchandise found, bought, bartered, or gained within his admiralty.

3.  Columbus or his lieutenant would be the sole judge in all disputes arising from traffic between those countries and Spain, provided the high admiral of Castile had similar jurisdiction in his district.

4.  Columbus might, at any time, contribute an eighth of the expense in fitting out vessels for this enterprise and receive an eighth of the profits. He fulfilled this last point by adding a third vessel to the fleet with the assistance of the Pinzons of Palos.

Thus, one-eighth of the expense of this grand expedition, undertaken by one of the most powerful nations, was borne by the individual who conceived it and risked his life on its success. The sovereigns signed the agreement on April 17, 1492, at Santa Fe in the Vega or plain of Granada. A letter of privilege or commission was issued to Columbus on the 30th.

In the letter, the dignities and prerogatives of the Viceroy and Governor were made hereditary in his family, and he and his heirs were authorized to prefix the title of Don to their names. While both sovereigns signed these documents, the separate Crown of Castile defrayed all the expenses. During her life, few people other than Castilians were permitted to establish themselves in the new territories.

---

An Extreme Leader maintains unwavering enthusiasm throughout the journey. The leader's energy fuels the involved teams, creating an air of confidence, excitement, and passion. This contagious enthusiasm drives the teams to execute the Big Vision with dedication and vigor.

---

### Columbus Gets His Ships

The port of Palos de Moguer in Andalusia was designated as the location for fitting out the armada. Due to prior misconduct, the port's inhabitants had been condemned by the Royal Council to serve the Crown for one year with two armed caravels. A royal order was signed on April 30th, 1492, commanding the authorities of Palos to have the two caravels ready for sea within ten days and to place them and their crews at Columbus's disposal, who was also empowered to procure and fit out a third vessel.

The crews of all three ships were to receive the ordinary wages of seamen employed in armed vessels and be paid four months in advance. They were to sail in the direction Columbus commanded under royal authority and obey him in all matters, with the stipulation that neither he nor they were to go to St. George la Mina, on the coast of Guinea, or any other of the recently discovered possessions of Portugal. A certificate of good conduct, signed by Columbus, would discharge their obligation to the Crown.

The sovereigns also issued orders to the public

authorities and people of all ranks and conditions in the maritime boards of Andalusia, commanding them to furnish supplies and assistance of all kinds at reasonable prices for fitting out the vessels. Penalties were proclaimed for anyone causing impediments. No duties were to be extracted, and all criminal processes against individuals engaged in the expedition were to be suspended during their absence and for two months after their return.

In a heartfelt gesture characteristic of her considerate nature, Isabella issued a letter-patent on May 8th appointing Columbus's son Diego as a page to Prince Juan, the heir-apparent, with an allowance for his support, an honor granted only to the sons of distinguished persons. After delays and disappointments that would have reduced an ordinary man to despair, Columbus left the court on May 12th for Palos.

Those who might be inclined to give up in the face of difficulties should remember that 18 years had elapsed from the time Columbus conceived his enterprise until he could carry it into effect. Most of that time was spent in hopeless solicitation amidst poverty, neglect, and ridicule. The prime of his life had wasted away in the struggle, and when his perseverance was finally crowned with success, he was about 56 years old. His example should encourage the enterprising never to despair.

Columbus had more than once presented himself at the gates of the convent of La Rabida, but now he appeared in triumph. The worthy friar received Columbus with open arms, and Columbus was once again the friar's guest. Juan Perez's character and station gave Columbus significant influence in the area, and he exerted his best support for the enterprise. Together, they went on May 23rd to the church of

St. George in Palos.

There, the notary public formally read the royal order in the presence of the alcaldes, regidors, and many inhabitants, and full compliance was promised. When the nature of the enterprise became known, horror prevailed as the inhabitants believed the ships were doomed to destruction. The owners of the vessels refused to furnish them for such a desperate service, and the boldest seamen shrank from the wild cruise into the unknown ocean.

The gossip of Palos circulated all the frightful tales and fables that ignorance and superstition could conjure to deter anyone from embarking on the expedition. The extreme dread with which it was regarded by a maritime community composed of some of the most adventurous navigators of the age is strong evidence of the bold nature of this undertaking.

Despite the royal order and the promises of the local magistrates, weeks passed without any progress. The worthy prior backed Columbus with all his influence and eloquence, but it was in vain. Not a single vessel was procured. On June 20th, the sovereigns issued more mandates, ordering the magistrates to press into service any vessels they deemed suitable belonging to Spanish subjects and to oblige the masters and crews to sail with Columbus.

Juan de Penalosa, an officer of the royal household, was sent to ensure compliance with the order. While occupied with the business, he received two hundred maravedis per day. He was to exact this payment from the disobedient and delinquent and other penalties expressed in the mandate. Columbus attempted to act on this order in Palos and the neighboring town of Moguer, with little success other than throwing the communities into confusion, altercations, and disturbances.

Eventually, Martin Alonzo Pinzon, a wealthy and

enterprising navigator, took an interest in the enterprise. His understanding with Columbus might have included a division of profits. Without Pinzon's assistance, outfitting the small fleet with the necessary armament might have been impossible. He and his brother, Vicente Yanez Pinzon, also a courageous and able navigator who later rose to distinction, possessed vessels and seamen in their employ. They also had relatives in Palos and Moguer and significant influence in the community. They likely provided the eighth share of the expense Columbus was bound to advance and furnished at least one of the ships, resolving to take command and sail in the expedition. The ships were ready for sea within a month of their engagement.

Columbus ended up with three small vessels. Two were caravels, light barques not superior to river or coasting craft of more modern days, depicted in old paintings as open without decks in the center and built up high at the prow and stern with forecastles and cabins for crew accommodations. Only one of the three was fully decked. Columbus considered the ships' small size advantageous for a discovery voyage, enabling him to run close to shore and enter shallow rivers and harbors. That such long and perilous expeditions into unknown seas were undertaken in vessels without decks and survived violent tempests remains among the most remarkable aspects of these daring voyages.

---

An Extreme Leader understands that the truly hard work begins once resources are secured. They know how to switch gears from salesperson to operator. Many leaders struggle with this transition, leading to failure. They can sell the idea but cannot effectively deliver it or choose and motivate a team who can.

---

## Ship-Owner Sabotage

During the equipping of the vessels, numerous troubles arose. The Pinta, along with its owner and crew, had been pressed into service by the magistrates under the sovereigns' arbitrary mandate—an example of the despotic authority exercised over commerce at the time. Respectable individuals were compelled to engage, with their ships and personnel, in what appeared to them to be a mad and desperate enterprise.

The owners of the Pinta, Gomez Rascon, and Christoval Quintero, displayed great reluctance towards the voyage, engaging in quarrels and contentions. They and their allies created obstacles to delay or sabotage the expedition. The caulkers performed their work carelessly and imperfectly, and when ordered to redo it, they absconded. Some seamen who had enlisted regretted their decision, were dissuaded by their families, or sought to retract their commitment. Others deserted and went into hiding.

Everything had to be accomplished using the harshest and most arbitrary measures, defying widespread prejudice and opposition. All difficulties were overcome by the beginning of August, and the vessels were ready for sea. Columbus hoisted his flag on the largest ship, the Santa Maria, which had decks. Martin Alonzo Pinzon, accompanied by his brother Francisco Martin as pilot, commanded the second ship, the Pinta. The third ship, the Nina, had lateen sails and was commanded by the third brother, Vicente Yanez Pinzon.

The expedition included three other pilots: Sancho Ruiz, Pedro Alonzo Nino, and Bartolomeo Roldan. Rodrigo Sanchez of Segovia was the inspector-general of the armament, and Diego de Arana, a native of Cordova, served as chief alguazil.

Rodrigo de Escobar acted as the royal notary and was tasked with recording official transactions. The crew included a physician, a surgeon, private adventurers, several servants, and ninety mariners, totaling 120 people.

Before departing, Columbus confessed to Friar Juan Perez and took communion, followed by his officers and crew. A deep gloom spread over Palos as they departed, with everyone having a relative on board. The sight of tears and dismal forebodings further depressed the spirits of the seamen beyond their existing fears.

An Extreme Leader remains vigilant for saboteurs along the way. Big Visions often attract detractors who want the project to fail to validate their skepticism. A leader identifies these individuals and their tendencies, taking decisive actions to separate them from the execution phases to ensure the project's success.

### Columbus Finally Departs on the Riskiest Venture

On the morning of August 3rd, 1492, Columbus set sail on his first voyage of discovery. Departing from a small island near Huelva, he steered southwesterly towards the Canary Islands, intending to strike due west then.

On the third day, the Pinta signaled distress: her rudder was broken and unhinged. Columbus suspected sabotage by the caravel's owners, Gomez Rascon and Christoval Quintero, who aimed to disable the vessel and cause her to be left behind. Despite this setback, Columbus navigated the Pinta to the Canary Islands for repairs, further delaying his journey.

This narrative is just a part of the story. Still, it highlights

Columbus's perseverance and vision, who accomplished one of the most incredible feats of all time despite almost two decades of struggle and occasional despair.

Columbus faced incredible obstacles on his voyage, including maintaining morale, avoiding mutiny, and sustaining his convictions in his Big Vision. Despite these challenges, he persevered and achieved his remarkable vision. Columbus was not an overnight success; he accomplished what no one had done before and what no one could ever claim again. He was the epitome of an extreme leader.

# DR. ELISHA KANE, CAPTAIN OF THE *ADVANCE*

A Case Study of the Essence of Extreme Leadership
in Extreme Circumstances

Based on the book

*Arctic Explorations in the Years 1853, '54, '55*

by

Dr. Elisha Kane

With excerpts, edits, paraphrasing, and
commentary

by

Charles Patton

# EXTREME LEADERSHIP

## Preamble

This full story of Dr. Kane and his men is available in fictionalized form in the author's book, Tigers of the Ice, available from Amazon. This novel is rooted in historical facts and uses the term 'Esquimaux' to accurately reflect the language of the time for Indigenous Arctic residents, now correctly referred to as 'Inuit'. This choice is not intended to uphold outdated or disrespectful views, and the depiction of Esquimaux characters, in that book as in this one, is respectful. The historical term "sledge" is also used, as in the original story, to refer to their dog sleds.

## Introduction

From as early as the year 1500, France, England, and the Netherlands began searching for a shorter sea route to Asia, one faster than sailing around the tips of Africa or South America. The most sought-after route was across the northern extremity of the Americas, through the waters north of Canada. This potential route became known as the Northwest Passage. For four hundred years, no one knew if those waters could be navigated to Asia. Many attempted the journey, but none succeeded until Roald Amundsen did so between 1904 and 1906. Countless explorers perished trying to chart these desolate waters.

Even in the mid-1800s, a significant misconception persisted among scientists-the belief that salt water did not freeze. This misunderstanding, coupled with the fact that icebergs were thought to be made of fresh water, led to the erroneous assumption that a sea route like the Northwest Passage would be open year-round. This misguided optimism drew the attention of Sir John Franklin, who in

1845 set out to chart as much of the Northwest Passage as possible. However, the harsh reality of these frozen waters claimed Sir Franklin, sparking a relentless quest to uncover his fate.

Dr. Elisha Kent Kane, a medical doctor, had served under Lt. De Haven on the Grinnell Expedition in 1850, which aimed to discover Sir Franklin's and his men's fate. Sir John Franklin, a rear admiral in the British Royal Navy, had vanished sometime between 1845 and 1847 with three ships, twenty-four officers, and 110 men in the Arctic while searching for the Northwest Passage. The first Grinnell Expedition failed to find any trace of Sir Franklin's party.

> An essential trait of an extreme leader is wisdom gained through experience. This wisdom need not come from extreme leadership conditions but must be relevant to the environment in which the extreme circumstances arise. In this case, Kane had previously participated in an Arctic expedition and understood the harsh conditions and demands it placed on men.

After the first attempt failed, Kane proposed and was approved, under special orders from the U.S. Navy, to lead the Second Grinnell Expedition in search of Sir John Franklin and his men. The expedition was named after Henry Grinnell, who provided the ship that Kane would captain, the Advance. The experiences of Kane, his officers, and his crew on the second expedition were extraordinary and extreme.

## Departure

In the spring of 1853, Kane departed on the Advance with eighteen officers, scientists, and crew. En route, he picked up a 19-year-old Esquimaux named Hans Cristian and several Newfoundland and Esquimaux sledge dogs. With this small crew, their sledge teams, and their 140-ton hermaphrodite brig, thoroughly tested in previous Arctic encounters, they sailed north along the western shore of Greenland. They had several dangerous encounters with icebergs but managed to sail far beyond civilization before being stopped by ice. In early September, they found shelter for their ship in the small Rensselaer Harbor, where it froze into the ice. With the sun ready to disappear below the horizon on October 24th, they had much preparation to do to survive the winter. They planned to search for Sir Franklin as weather permitted during the winter and in earnest come early spring.

To survive the winter, they adopted the lifestyle of sailors, banking the ship's sides with snow to conserve heat and stretch their coal supply. They spent any spare moments sending out hunting parties, smoking out rats (nearly setting the ship on fire in the process), and operating an observatory for scientific observations, which was also part of their mission.

## Their Situation Worsens

But then trouble began. Their dog team, critical to their survival, started losing members to an unknown "rabies-like" disease. Hunting for walrus, seals, hares, and birds became increasingly difficult and less successful as winter progressed until only Hans had any success, but that was rare. Fresh meat was their most important defense against scurvy.

Occasionally, a hunting or search party would get caught on the ice when it began to crack and move. Although temperatures were 25 to 50 degrees below zero, they maintained 60 to 65 degrees inside the ship. However, limited movement due to the temperature led to profound boredom. They built a false bulkhead to reduce the area heated by their stove, creating crowded but efficient conditions. Venting the confined area also posed problems with fumes, soot, and the risk of carbon monoxide poisoning.

By January, temperatures consistently ranged between 60 to 75 degrees below zero, which is more than one hundred degrees below freezing (32 degrees above zero). All the chemicals used for scientific observations froze at these temperatures, including oils, naphtha, and chloroform. From October through March, they saw no sign of the sun. The constant darkness caused dismay among the men and surviving dogs, and a strange insomnia set in. Puppies were born but also succumbed to the disease. Their fuel supply, coal, depleted faster than planned, forcing them to ration it to three buckets per day. They relied on dried foods and bread with no fresh meat and only a barrel of potatoes left. All but two men showed severe scurvy effects, including loss of energy, loosening teeth, receding gums, swollen legs and feet, and pallor.

By March, temperatures remained 40 to 50 degrees below zero, but the sun began showing at 15 degrees above the horizon for short portions of the day. The spring tides raised and lowered the ship, still trapped in the ice, seventeen feet twice per day. Attempts to move the ship using harnesses for men, called Rue-Raddies, failed. In late March, as temperatures warmed to 40 and 20 below, they resumed their search for Franklin in earnest. More than once, the search party faced trouble from frostbite, exhaustion, and scurvy, requiring rescue themselves. Men staggering and collapsing in the snow, and delirium and weakness from their sparse diet became all too common.

By April, one man, Jefferson Baker, had died. The day before Baker died, he was seized by lockjaw and emitted the most ominous and frightful sounds the ship's physician, Dr. Hayes, had ever heard. As Baker neared death, a man on deck watch reported men approaching the ship. The group that approached made wild gesticulations, tossing their heads and arms, but appeared unarmed. They were native Esquimaux and spoke no English. One of Kane's men, Petersen, was brought from his sick bed to interpret some of their language.

> Like everyone else, leaders carry biases toward certain people and their processes. These biases are not about racial stereotyping but are based on perceived differences in approaches, attitudes, and intelligence. In this regard, Kane formed biases about the uncivilized native Esquimaux, distinct from his relationship with Hans, a more civilized native. However, a good leader keeps an open mind. By doing so, Kane learned important survival lessons from the Esquimaux.

After a private interview onboard with their leader, Metek, Kane allowed the others inside the ship. Initially, Kane feared them, but he gradually learned that their innocent behavior made them more of a nuisance than a threat. He found them to be a happy group. He discovered that they did not comprehend the idea of possessions or a future. As a result, they would rob the ship as readily as they would let Kane use their dog teams and sledges. They would consume all the food available on the day they got it without conserving any for the next day, embodying the concept of "living in the moment."

Over the next few months, while searching for Franklin, Kane observed how the Esquimaux lived, dressed, hunted, and survived in such a harsh climate. Kane's carpenter, Ohlsen, learned to make lighter and stronger sledges by emulating Esquimaux designs. The ship's crew joined Esquimaux hunting parties, finding them far more effective. Kane and the others adopted the Esquimaux way of dressing, in furs from head to toe.

As the search for Franklin continued, the expedition faced numerous challenges. The constant sunlight led to snow-blindness, and the thinning ice made sledging more perilous. On one occasion, Kane's sledge and dogs plunged through the ice, putting his life in jeopardy. Despite these hardships, the crew persisted in their search, their determination unwavering even in the face of frustration at not finding any sign of Franklin's party.

Handling dogs pulling a sledge with their leads constantly tangled was particularly difficult. The dogs fought among themselves and using an 18-foot-long whip to steer them was challenging. Additionally, the ice often broke up and compressed into 10- to 20-foot-high vertical blocks that could be traversed only with axes and great exertion. Even with increased fresh meat and some greens, the men continued to suffer from scurvy. By June, 11 men were sick with scurvy and snow blindness. One man, Brooks, had to have his foot amputated due to scurvy. On June 5th, Kane sent a search party southwest in one last effort to find Franklin before departing, as the ice was due to melt soon. Hans, who remained healthy, brought seals daily to sustain the men. Petersen also harvested handfuls of one-inch-high "scurvy grass" to feed the men.

When the search party returned unsuccessfully, one of the men, McGary, was blind from the sun. He reported to Kane that a polar bear had stuck its head into their tent while sleeping in a tent on the ice with their guns on the sledge. They managed to fend it off but failed to frighten it away using burning papers and matches. One of the men, Hickey, cut a hole in the back of the tent, fetched a boat hook, and managed to beat the bear back beyond the sled, where he secured a gun and, firing it, scared off the bear.

As the nearby ice was no longer safe for travel but still held their ship tightly, their searches were limited in range and consistently unsuccessful. The summer wore on, but the ice did not break as normally. Kane became concerned about their ability to leave as planned. They tried blasting the ice around them to create an opening (lead) in the ice. Their hopes hinged on the late August and early September gales to free their ship, as they were ill-prepared in health, food, and fuel to last another winter.

### Their Situation Becomes Desperate

By mid-August, Kane realized they would not escape the ice this year and would have to endure a harsher winter than before. As a result, he reduced fuel use to six pounds of wood per day to conserve what little coal remained. This meager amount allowed for coffee twice a day and soup once. The prospect of another year of disease and darkness without fresh food and fuel was described by Kane as "horrible to think of." They managed to free the ship enough to warp (i.e., move by pulling with ropes) her inside a group of small islands to what seemed like a better berth. Then, Kane realized they would have to face the impending winter head-on. Despite the uncertainty, he knew that inaction was not an option.

> An extreme leader anticipates as much as possible, plans, expects the best while preparing for the worst, and conserves resources for harder times. These are essential practices for anyone who might face extreme challenges. Being prepared and in the right place at the right time is often seen as luck, and indeed, luck can play a crucial role in achieving success in extreme situations.

At this point, a group of men believed that escaping on foot to the south was still feasible. Kane faced a difficult decision. He was convinced they were wrong and that such an attempt would be dangerous. However, he could not expect his associates to follow his conclusions because, according to nautical rules, when a ship is hopelessly beset, the master's authority gives way, and the crew is permitted to take their counsel. None of them had signed up for a cruise that would last two winters on a ship stuck in the Arctic ice.

### The Crisis Point

Kane believed their safety would be more secure with a half-dozen resolute men remaining on the ship than if they left this late in the season. He called the officers and men together, explained how he viewed the situation, reviewed the hazards they would face trying to trek out, and gave them 24 hours to deliberate. He told them that at the end of that time, anyone still wanting to leave and willing to provide a written explanation of their reasons would receive the best outfit he could provide, a more than abundant share of the remaining stores, and his goodbye blessing.

At noon the next day, Kane further endeavored to show them that an escape to open water would not succeed, that they had duties to the ship and the other men, and strenuously asked them to forgo their plans. He insisted that those who left would need to put themselves under the command of the officers who might choose to join them and that they would need to renounce in writing any claims against those who chose to remain. Kane then took roll, and eight of the seventeen survivors resolved to stay, including Hans the Esquimaux. Dr. Hayes volunteered to accompany the men who would leave, believing they would need a doctor. Kane divided the stores liberally and saw them off as promised on August 28th with a written assurance that they would receive a brother's welcome if they were driven back. One man, Riley, returned in a few days.

---

### Kane's Skills as an Extreme Leader

1. He addressed the problem promptly but not hastily, understanding that timing is crucial.

2. He communicated a clear plan effectively to those who would stick with him.

3. Rather than trying to hold back dissenters, he allowed them to leave, thereby removing negative influences from those who supported his vision and goals.

4. He solidified his position with those who remained, immediately putting them to work on the plan to focus on the positive and reduce the impact of negativity.

The skills Kane applied to this challenge are essential for extreme leaders:

### Adapt or Die

With the remaining men now driven to the wall, their energies were quickened, not depressed. Kane reviewed his plans with them to ensure agreement and then set them to work. He believed that systematic action was essential to combat difficulties and improve morale. He laid out duties, schedules, and assignments.

Learning from the Esquimaux, Kane implemented their living, eating, and dressing practices. He transformed an 18-square-foot area of the interior of the brig into an igloo, packing turf and moss from floor to ceiling for insulation. This smaller area would be warmed by their body heat, reducing the need for fuel. The floor was caulked with plaster of Paris and ordinary paste, covered with several inches of Manila oakum, and topped with a canvas carpet. He created an igloo-like entrance with a low moss-lined tunnel, doors, and curtains. Despite the hard work, exacerbated by scurvy and dropping temperatures, the crew continued "bravely on," gathering moss to cover their roof and stripping off the outer deck planking for firewood.

By September, a year since they were first frozen into the ice, the wild game began disappearing. An attempt to hunt seals resulted in Kane crashing through the ice and nearly drowning. He had to abandon his sledge, kayak, tent, guns, snowshoes, and everything else he had brought. To avoid freezing to death, he ran twelve miles back to the brig with Hans "frictioning" him as he ran.

Problems with the native Esquimaux persisted. Kane considered what the Esquimaux saw as "sharing" as theft, especially when articles like cooking vessels and lamps were critical to survival. Kane decided to send a message to the Esquimaux elders. He dispatched two men, Morton and Reilly, to overtake the thieves and bring them back to the brig for punishment. Despite the 30-mile trek each way, no one complained. He placed the thieves—Myouk, his wife Sievu, and Aningna, wife of Marsinga—in the brig's hold and sent Myouk with a message to their chief, Metek, at the village of Etah, demanding a ransom.

> An extreme leader knows when to draw the line. While understanding, cooperation, and accommodation are important, there are times when a stand must be taken, regardless of the potential risks. An extreme leader recognizes where that line is and has a plan to address those who cross it too often.

After five days, the chief arrived with another elder, Ootuniah, bringing a sledge-load of misappropriated items: knives, tin cups, scraps of wood and iron, and other stolen goods. Having not seen guns before the ship's arrival, the natives had become afraid of their "fire-death." Metek admitted to Kane that their "strength" was leaving them when the contingent of men went south. Kane proposed a treaty, and Metek agreed they would no longer steal, bring them fresh meat, sell or lend them dogs whenever available, and show them where to find game. In return, Kane promised not to use the "fire-death" or sorcery against them and welcomed them onboard the brig. To seal the deal, he gave Metek gifts, including needles, pins, knives, wood, fat, an awl, and sewing thread. In exchange, Metek gave Kane high-quality walrus and seal meat. This treaty was never broken.

From that point forward, their dogs were considered common property, and the Esquimaux shared food even during times of starvation. They provided critical supplies of meat and became close allies and friends.

Throughout the winter, the healthier men made long treks up to 160 miles with dogs along the flat ice near the shore to hunt. They also raised the Advance above the flotation line to prevent it from being crushed by the ice at spring low tides. Kane and the carpenter, Ohlsen, calculated they could burn seven or eight tons of fuel cut from the brig without compromising her seaworthiness. Kane estimated they could make it through if they limited fuel consumption to seventy pounds daily. They cleaned and reconfigured the stove's smoke tubes and ice-melting pans into a more efficient system for heating and melting ice.

In late October, Erebus, one of the black dogs given to the men who had left, returned to the brig, having broken loose. Hans and Morton tracked the Esquimaux to the lower village of Etah and brought back 270 pounds of walrus and a couple of foxes, which, along with the remaining meat of two bears, would have to last until daylight returned. The lack of fresh meat caused severe wasting among the men. Wilson, Brooks, Morton, and Hans were bedridden, leaving only six men to do all the work. While Hans recovered, Goodfellow fell ill.

Kane was in awe of the Esquimaux's profound knowledge of the environment. Their ability to predict bird migrations, understand animal habits, and identify open water ranges was truly remarkable. They could even find liquid fresh water in the frozen darkness by tapping the ice with a pole and listening.

In late November, Hans and Kane were checking traps in the dark when a polar bear caught their scent. They were unarmed and could hear the bear close by. Standing on the ice foot about ten feet above the flow ice, Kane instructed Hans to run for the ship while he played decoy. Kane lay quietly, listening for the bear, until he saw a hummock move, roar, and charge. He ran as fast as he could, throwing off his mittens to distract the bear with their scent. He reached the brig where Hans was already advancing with Kane's rifle. They retraced Kane's footprints but found only one glove and no bear.

In early December, Esquimaux arrived with five sledges carrying two men who had left the brig months before. Bonsall and Petersen had left the rest of their party seventy miles away, "divided in counsel, their energies broken, and their provisions nearly gone." Only Kane, McGary, and Hans were fit to travel, so Kane organized a rescue party. He packed 350 pounds of food and some tea, dispatching the supplies with the Esquimaux, although he was reluctant to trust them entirely. He prepared Hans to follow their track the next day to ensure the supplies reached their destination.

In just four days, the Esquimaux returned with the men. With temperatures at 50 degrees below zero, the men arrived covered in rime and snow, fainting from hunger. Kane welcomed them back as promised, sharing meat-biscuit soup, molasses, wheat bread, and salt pork. They had traveled 350 miles, surviving on limited frozen seal and walrus meat.

An extreme leader does not hold grudges but maintains an open mind. At the same time, an extreme leader is cautious not to be taken advantage of and can distinguish honest repentance from manipulation and deceit. This skill is where leadership wisdom comes into play. Experience with good and bad people builds the ability to discern who should be forgiven and who should be driven away. While we are not on earth to judge worthiness before God, we must judge others when allowing them into our lives.

Kane learned from the Esquimaux that his men had appropriated clothing, fox skins, and other items using their superior force while traveling and stopping at huts along the way. Given the pact he had formed with the Esquimaux, Kane saw this as disrespectful and improper. He held an informal hearing and returned the items with five needles, a file, and a wood stick, which satisfied the Esquimaux. Kane then fed everyone a large dinner, and all fell asleep around the stove.

Due to the crowding's impact on air quality in their tiny room, Kane moved four lamps outside and assigned a watch. However, the watch fell asleep, and the lamps set the walls, bulkhead, dry timbers, and even the brig's skin on fire. Fortunately, Kane had kept a hole open in the ice for emergencies. By passing buckets of water and using furs brought by the returning men, they extinguished the fire before the entire room and ship burned. Kane collapsed from smoke and steam inhalation and, when carried on deck, was found to have lost his beard, eyebrows, and forelock, with burns on his forehead and palms. The transition from the fire to the outside, at 46 degrees below zero, was extremely challenging, and nearly every man suffered frostbitten fingers or toes.

Christmas passed with a modest celebration, and the New Year, 1855, arrived. With twelve lamps constantly going for heat, everything became covered in grease and soot. The Esquimaux eventually left, and concerned about the air quality, Kane reduced the number of lamps to four, with two vented to the chimney. By this time, Kane had lost over fifty dogs to a disease. The only food available for the remaining five dogs was a concoction made from the carcasses of the dead dogs. Only one Newfoundland and three Esquimaux dogs remained of the original fifty. Kane dreaded the thought of a 100-mile run with dogs that could drop at any moment, but he planned to do it in hopes of obtaining desperately needed food.

Darkness continued for months, with 140 days of total darkness each year in this area of the Arctic. The men had adapted almost entirely to the ways of the Esquimaux, preferring raw blubber, frozen walrus meat, and walrus liver for their taste and ascorbic benefits against scurvy.

However, Wilson's health declined, and only three men remained healthy: Kane, Hans, and Ohlsen. Kane decided it was time for him and Hans to attempt the 93-mile walk in bitter cold and total darkness. He had no choice but to go, but the weather turned severe, forcing him to wait. The temperature hovered between minus 60 and minus 40, with severe gale winds. Kane prepared a lightweight sledge, weighing only forty pounds and flexible like a "lady's work-basket," with minimal iron to prevent brittleness in the cold. With only four serviceable dogs, the load had to be light, and he and Hans would jog alongside.

Kane dressed as the Esquimaux did, with a fox-skin jumper called a kapetah featuring an air-tight hood. Beneath it, he wore a bird-skin shirt, softened by women chewing the skins, with the down next to the body. As many as five hundred auks contributed to a single shirt. His britches were made from polar bear skin, and his foot gear consisted of bird-skin socks with grass padding, bear skin, and straw. The clothing was loose around the waist and open to the atmosphere below. An Esquimaux could sleep on a sledge at 93 degrees below zero without consequences in this dress. Additional articles included a fox tail held between the teeth to protect the nose from wind and sealskin mitts wadded with straw. Kane observed that "our party of American Hyperboreans are mere carpet-knights compared to these indomitable savages." He added layers of wool and fur beneath his outfit, without which he could not survive below minus 50 degrees.

Extreme leaders adapt rapidly to their surroundings, learning lessons quickly in an emergency and maintaining flexibility in changing conditions. Creativity in responding to these changes is essential, and a leader with trusted advisors is more likely to achieve the necessary innovation in a crisis. However, caution is needed to avoid procrastination in resolving differences of opinion among advisors. Extreme leaders must use their wits and wisdom to keep actions moving in the right direction and on time.

The ice grew so thick that it ground against the ship's bottom at low tide, causing dangerous upheaval. Ohlsen, the carpenter, reported that the cross beams were bent six inches, indicating a perilous pressure level. Any leakage would be disastrous in their condition. By late January, Kane knew March would be their critical month. To stretch their supply of firewood, they began burning pieces of tar-laid hemp hawsers and reduced their wood consumption to thirty-nine pounds per day. Scurvy symptoms worsened, with swelling limbs, retracting gums, hemorrhaging, and severe despondency. Dr. Hayes had to amputate his toes due to frostbite. Only five men remained able to work: Ohlsen, Bonsall, Petersen, Hans, and Kane. By the end of January, the weather still had not permitted Kane and Hans to leave for the village of Etah to barter for food.

> Effective extreme leaders prioritize the well-being of their supporters, even sacrificing their own needs and comforts to ensure their teams are met first. They share any glory and success with their followers, but in the face of failure, they take full responsibility and never place blame on their team.

Finally, on February 4th, Kane sent Petersen and Hans south. Ohlsen had collapsed, leaving only Kane and Bonsall able to stand. Kane had to give up trapping due to a lack of meat for bait. Any water settling below two feet off the floor would freeze in their cabin. Four days later, Petersen and Hans returned, unable to reach Etah because Petersen had broken down with scurvy. Their return allowed Hans and Kane to hunt, and they killed two rabbits, providing the men with their first fresh meat in ten days. By mid-February, Hans's rabbit-hunting success improved. Kane knew they would all be lost if Hans's health failed. Kane felt optimistic that Hans' continued hunting would help contain the scurvy. He also believed that the returning sun would boost morale and that they would stick together as long as they stayed alive. A week later, Hans shot a deer, yielding 180 pounds of meat, but unfortunately, it putrefied quickly, and most was lost.

By early March, the sun began to peek above the horizon. Kane reluctantly sent Hans to the village of Etah on the 93-mile trip, expecting him back in four to five days. Hans returned in four days, reporting that the settlement was in a state of famine; they had even eaten twenty-six of their thirty dogs. Hans managed to shoot a walrus while there and returned with his share of the meat and an Esquimaux, Myouk, whom Kane had asked Hans to bring to help with the hunting.

## Desertion Cannot Be Tolerated

By mid-March, they had burned the last Manila hawser, so Kane had Petersen strip more wood from the brig. Kane also learned of a desertion by two men planning to steal Hans's sledge and dogs. Such an act would be fatal for the sick men, so Kane had to address it. He confronted the plotters, stopped the plan, and punished the perpetrators. However, one of them, Godfrey, escaped and headed south alone. Kane worried Godfrey would reach Etah and steal Hans's sledge and dog team. Kane would have pursued him immediately but could not leave due to the sick men's constant need for food.

Extreme leaders take whatever action is necessary to protect their followers. Threats to their health and well-being must be dealt with firmly and promptly, regardless of the risk. Underminers must be removed as quickly as they are discovered, even at the expense of short-term progress.

188

By late March, Kane could not help but compare their current condition to a year before. Although they had adapted well, the men's health had significantly deteriorated. Cutting away two days' worth of fuel from the brig took them a whole day. Godfrey and the potential danger he posed to Hans weighed heavily on Kane's mind, but the condition of his men prevented him from acting.

On April 2nd, a man was spotted near the brig, and Kane went to investigate, leaving Bonsall armed on the deck. He located the man among the hummocks; it was Godfrey. Seeing Kane unarmed, Godfrey allowed him to approach, unaware that Kane had a concealed pistol. Kane drew it and forced Godfrey back to the ship, but Godfrey refused to board. Kane left him under Bonsall's guard and went below for irons. As soon as he reappeared, Godfrey bolted. Bonsall's weapon misfired, and Kane's first rifle shot went off accidentally, while a second aimed shot missed. Godfrey escaped, but Kane discovered the sledge and dogs Godfrey had stolen, along with walrus meat, which was a godsend. Godfrey had inadvertently done more good than harm.

Kane now firmly believed Godfrey was dangerous. Hans had been gone for over two weeks, much longer than the usual four to five days. Kane ordered that any further act of desertion would be met with the sternest penalty, which would be death. Kane's concern for Hans grew, knowing someone would have to search for his trusted friend, and it would have to be him.

On April 10th, Kane left with five dogs, the smallest sledge, and the lightest load possible. Describing himself as "more than half Esquimaux," he took only an extra jumper, sack pants for sleeping, and a frozen ball of raw walrus meat packed with tallow. After covering sixty-four miles in 11 hours, Kane saw a speck on the ice, recognizing Hans by his gait. They reunited, chattering in a mix of Esquimaux and English. Hans had fallen ill and was down for five days. They settled in a broken-down stone hut at Anoatuk, where they enjoyed tea and molasses that Kane had brought, knowing it was Hans's favorite. Hans carried a few lumps of walrus liver, which they shared. Hans had hunted successfully with the Etah Esquimaux and stashed his share of the meat on a nearby island. He had been nursed back to health by Shunghu, a young daughter of one of the elders, who might have touched his heart.

Hans told Kane how Godfrey had tried to persuade him to travel south with him and, when refused, tried to steal his rifle. Hans overpowered him, and Godfrey proposed taking some of Hans's walrus meat back to the ship to make terms with Kane. Kane believed Godfrey intended to link up with another man who had planned to follow but was detained. Kane decided Godfrey needed to be captured and confined on the brig for everyone's safety.

Kane sent Hans to a lower Esquimaux settlement near Cape Alexander to negotiate for four dogs, offering his remaining dog team when they headed south. Kane realized that if the ice did not melt, he must be prepared to move south without the ship and decided to make a final decision in early summer.

On April 12th, Kane returned to the brig. The men were without fresh food until Hans arrived with rabbits and walrus liver. Kane had become more like the Esquimaux, adapting their attitude of living day by day. Hans failed to negotiate for additional dogs but brought back two Esquimaux, Metek and his nephew, to help with hunting. Metek reported that the winter had been hard on the Esquimaux, with many dogs killed to survive. Kane estimated that the eight known settlements, totaling about 140 members, had no more than twenty dogs remaining. Natural deaths, three murders, and infanticide had further depleted their numbers. The tribes laughed when Kane explained they were progressing toward extinction.

Kane devised a plan to recapture Godfrey. He traded places with Metek's nephew, dressed like him, and accompanied Metek back to Etah. His trick worked; when the villagers turned out to greet Metek, Godfrey was among them. Kane got close to him and placed a pistol to his ear before Godfrey knew he was there. Kane put Godfrey in irons and transported him back to the brig.

A week later, Kane sent Hans on another mission to borrow dogs, sending an iron bar suitable for making harpoon shafts to aid the transaction. Hans returned with walrus meat and three Esquimaux, each with a sledge and dog team equipped for a hunt. Kane described the party leader, Kalutunah, as a noble savage. Kane proposed one last expedition as far north as possible.

Within days, the three Esquimaux, Kane, and Hans departed. They traveled until early May, finding no sign of Sir Franklin but discovering new valleys, capes, and harbors, documenting vistas never seen by non-Esquimaux. Kane observed the Esquimaux attacking and killing a polar bear with lances and knives, noting that five out of seven Esquimaux bore scars from bear attacks.

Preparations for departure had been ongoing for nine months, but after Kane returned from his final search, efforts intensified.

When a plan fails, and a radical change in direction is needed, an extreme leader does not hesitate to make the change. Solid planning for the alternative is essential, as a second failure would devastate morale and erode followers' support. Even if the initial failure is not the leader's or followers' fault, avoiding a second failure is crucial. Followers may understand one failure, but a second is far less tolerable.

### Abandoning Ship is Never Easy

It was clear to Kane that the brig could not be saved and would need to be abandoned. His primary plan, laid out to his men, involved loading the brig's three boats with only the most essential equipment, food supplies, and the sick, then pushing and pulling them across jumbled and dangerous ice. This ice was filled with tall hummocks, miles of cracks, and subjected to gale-force storms to the closest open water, hundreds of miles to the south. To facilitate movement, they would attach runners to the bottom of the three boats— *Faith, Hope*, and, oddly, the *Red Boat*.

For food, they took ship's bread pounded into powder, pork fat, and tallow melted down and frozen, concentrated bean soup frozen and bagged, and the remaining supply of flour and meat biscuit sewn into double bags to protect from moisture. They planned to hunt along the way for additional meat, a testament to their resourcefulness and determination to survive.

They held a solemn memorial upon leaving the ship, a moment that underscored the gravity of their situation. During this ceremony, all men signed a document placing them under Kane's command. Kane also signed a letter attesting that the brig could not be saved. Due to dwindling provisions, the mission to find Sir Franklin could no longer be pursued, and further cutting of firewood would render the ship unseaworthy. Wintering in the brig for another year was not an option. The weight of these decisions was intense.

As their travel across the ice was slow and tedious, many trips had to be made back to the brig to cook and secure additional items.

> An extreme leader recognizes and honors the sacrifices of others at every opportunity. Even the abandonment of a failed effort can be an occasion for celebration. Kane viewed the Advance as a family member, having sheltered and protected them through two harsh winters and giving up her life for their survival. She deserved their respect as they left her behind.

Putting every man available to the task, Kane assigned six men to move each boat. With only twelve healthy men, they could move only two of the three boats at a time. Kane anticipated this and set up a "field hospital" in the hut at Anoatuk, where the ill could rest and avoid much of the rugged travel. He then transported each of the four ill men to the hut, saving them from exposure to the elements. Kane believed the hut saved their lives by giving them time to recover and a change of scenery to improve their morale. By this time, Godfrey had been reincorporated into the work team, displaying an acceptable degree of trustworthiness.

Kane dispatched Godfrey to Etah to obtain meat. Godfrey returned with the meat and brought Metek, who helped with the task. Despite their efforts, after 14 hours, the weakened men had moved the boats over twelve miles of surface but advanced only a mile and a half from their starting point, the brig.

Kane repeatedly returned to the brig to bake bread, miraculously raising it in three hours without baking soda, salt, or shortening. Kane and his men moved 1,500 pounds of provisions from the brig to the boats. While this may sound like a lot, their heavy exertion, including the dogs, consumed more than one hundred pounds per day.

When Kane began moving provisions to Etah, he found the ice turning sodden and no longer trustworthy. He worried about losing critical supplies. When he reached Etah, he saw the inhabitants enjoying a feast of auks, which had returned in great numbers. They seemed unconcerned about the following winter, eating what they caught. Kane exchanged dog teams with them, as theirs were now well-fed. An elder provided Kane with a sledge of walrus meat and sent two young men to assist him through a nearby section of broken ice. Before leaving, Kane marveled at the children's carefree play and their ordinary life in such an extraordinary place. It pained him to think of the winters they endured year after year. The families lived as one big family, with villages spaced a day's dog-march apart, where hunting was good. Their routes were so familiar that the dogs knew where to go without guidance. The Esquimaux had named every change in season, every rock, every plant, and every safe place to cache meat. Kane believed that, while they had no particular resistance to exposure and fatigue, they had vast knowledge of the dangers around them and knew how to deal with threats that would daunt non-Esquimaux.

An extreme leader works as hard, if not harder, and sacrifices as much, if not more, than what they ask of others. This standard is the mark of a true and extreme leader. Leaders do not place themselves above those who follow. This humility and dedication earn the respect needed to maintain their position of authority.

When Kane returned to the boats, a strong storm had driven the men to turn the boats over and crawl under them for shelter. Despite their dire circumstances, their appetites remained strong, each man consuming more than three pounds of food daily. Kane returned to Etah with Petersen to ask for more assistance from Esquimaux in moving the boats. They departed amidst a fierce cyclone-force gale that Kane, Petersen, and the dogs had to lie flat on the ice to avoid being blown away. At times, they lifted the sledge on their shoulders and ran to the shelter of a small island. Exhausted, they reached firm ground, safe from the ice's unpredictable movements but not out of danger. The storm's intensity made it as dark as night despite being full daytime, and they struggled to see each other or the dogs. Finding no refuge on the island, Kane decided they had to make for the mainland across the dangerous ice.

Upon reaching the shore, they found a 30-foot-high rock with drifts piled against it. Barely able to dig a burrow, they dragged the dogs into the snow drifts. More snow piled over them as the storm raged. Their fur jumpers had been blown off and soon wet to the skin. A dog fight broke out, causing their canopy to collapse and exposing them to the elements. The noise and tumult of the storm were unlike anything Kane had experienced at sea. Snow drifted over them again, and they waited. Kane used the storm to their advantage, returning to the boats with the wind at their backs. They reached the boats in 20 hours, covering only forty miles.

They suffered several accidents, with the boats falling through the ice. In one incident, the Faith was nearly lost. Ohlsen, a powerful man, used a capstan bar to save the sinking boat but lost his balance, injuring himself fatally. His heroic efforts saved the boat and potentially his comrades' lives.

At one point, they raised sails on the boats and crossed a stretch of flat ice, covering a distance equal to the previous five in one day. Soon, leaders from Etah arrived with Kane's dogs, who had now recovered. Kane dispatched the dogs with a sledge to retrieve the sick from Anoatuk and bring them to Etah. Five men and two women from Etah arrived to help a week later. Despite the added help, the boats periodically broke through the ice.

Kane had not seen Hans for two months. Hans had asked to go south to replace his boots and get walrus hide for soles, promising to walk to avoid tying up the dogs. Kane expected him back in three to four weeks but soon realized he had been missing for too long. Inquiries revealed that Hans had stopped by Etah, arranged for the leather, and then traveled to Peteravik, where Shanghu and his daughter lived. Kane never saw him again but heard he had moved south and married.

Another accident almost lost the *Red Boat* and all the voyage documents. Morton saved the boat through extreme efforts, with Bonsall pulling him from the icy water by his hair.

They finally reached Etah, enjoyed a brief celebration and a feast of auks, and then pushed their boats to the ice's edge, where open water extended to the horizon. They said their goodbyes, launched their boats and sailed into the treacherous, iceberg-filled sea. They sailed, rowed, and got stuck numerous times in the ice. A final chance shot with their final bullet secured a seal vital to staving off starvation. After 84 days at sea in small boats, they reached Upernavik, a Danish colony, their first encounter with civilization in three years.

The men struggled to adjust to indoor heat, preferring to sit outside on piles of snow during a reception in their honor. Kane also brought back two dogs, which he had carried as "meat on the hoof" but had grown too attached to kill. One dog, Toodla, had saved his life during an ice fall, making her especially dear to him.

Kane returned with fifteen of the eighteen men who had ventured north with him, an admirable outcome given their ordeal. The fate of Sir Franklin and his men remains a mystery, with only some evidence of encampments, sunken ships, and a few graves found. The whole story of their disappearance has never been discovered.

# MAHATMA GANDHI

A Case Study of an Extreme Leader with
Commitment, Determination and Persistence

Based on the Autobiography

*The Story of My Experiments with Truth*

by

Mahatma Gandhi and other

sources

With excerpts, edits, paraphrasing, and

commentary

by

Charles Patton

EXTREME LEADERSHIP

### His Early Life

Born on October 2nd, 1869, in Porbandar, India, Mohandas Karamchand Gandhi became one of the most respected spiritual and political leaders of the 20th century. He was instrumental in freeing the Indian people from British rule through the novel approach of nonviolent resistance. Gandhi was honored by Indians as the father of the Indian nation and was affectionately called Mahatma, meaning "Great Soul," and later Gandhiji as a mark of reverence. At age 13, he was married through a parental arrangement to a girl of the same age, and they had four children. Gandhi studied law in London and returned to India in 1891 to practice.

In 1893, he accepted a one-year contract to do legal work in South Africa, which was then under British control. There, he faced discrimination and abuse when he attempted to claim his rights as a British subject and saw that all Indians suffered similarly. This experience in South Africa was a turning point in Gandhi's life, as he spent 21 years there, working tirelessly to secure rights for Indian people, laying the foundation for his later work in India. (*Indian Child*, p. 1).

### His Big Vision

He led the campaign for Indian independence from Britain based on the principles of courage, nonviolence, and truth, a method of civil disobedience he called satyagraha— meaning "holding onto truth" and involving the withdrawal of cooperation from the state. Gandhi was arrested many times by the British for his activities, both in South Africa and India, and believed it honorable to go to jail for a righteous cause. Over his lifetime, he spent a total of seven years in prison. He often fasted to emphasize the importance of nonviolence.

Gandhi's philosophy was a catalyst for change, inspiring large masses of followers to boycott British educational institutions, law courts, and products. They resigned from government employment, refused to pay taxes, and abandoned British titles and honors. His aim was not just to achieve independence, but to have both oppressors and the oppressed recognize their shared humanity and common bonds, a testament to the power of his ideas.

Gandhi's grand vision was that civil disobedience could be a powerful tool for changing or eliminating morally wrong laws without resorting to violence.

### His Challenges

His strategy was not without setbacks.

In 1922, because of atrocities committed by his supporters against the police, he was forced to call off a campaign protesting against the British occupancy. He was imprisoned and remained so until 1925.

After his release from prison, Gandhi established a commune and a newspaper and initiated a series of reforms aimed at helping the socially disadvantaged, the rural poor, and the untouchables. He continued these efforts until 1930.

In 1930, he emerged again to lead a 248-mile march in protest against the British taxes on salt. He and thousands of followers illegally but symbolically made their salt from seawater. This defiance reflected India's determination to be free despite the imprisonment of thousands of protesters. As a result of this defiance, the Indian National Congress struggled for the next five years to put into law a complex agreement representing all the people of India. In February 1937, provincial autonomy became a reality.

India was granted independence in 1947. Gandhi had advocated for peaceful coexistence between Hindus and Muslims, but when the British partitioned India and Pakistan, violence broke out between the two religious groups. On January 13th, 1948, at age 78, he began a fast to stop the bloodshed between the two factions. After five days, the opposing leaders pledged to stop the fighting, and Gandhi broke his fast. Twelve days later, a Hindu fanatic who opposed his tolerance of all creeds and religions assassinated him.

### His Qualifications as an Extreme Leader.

Gandhi can be considered an extreme leader on many levels.

> What made Gandhi a successful leader? His methods amazed people and endeared him to his followers. The purity of his morals allowed people to listen to his message truly. His actions demonstrated the path to achieving his vision.

Throughout his life, Gandhi was a man of meager means. He did not require fancy surroundings and chose the opposite extreme. He epitomized the characteristics of an extreme leader by taking care of others before himself.

Gandhi took up weaving, traditionally seen as women's work, to symbolically embrace women's rights and weave those rights into the fabric of society by crossing gender lines. Gandhi was unimpressed by titles or social standing, focusing instead on each person's simple humanity. He performed his work on the borders, bridging gaps between people— England and India, Muslims, and Hindus, rich and poor.

> Boundaries play a crucial role in extreme leadership. Extreme leaders often operate on the frontiers of discoveries, resolve impossible disputes, and break down the arbitrary barriers that humans erect to protect vested interests.

Another characteristic of an extreme leader that Gandhi displayed was his humility. Although he appeared perfect to many, he never considered himself as such. Gandhi clearly understood the reward he sought. While some seek fame and fortune, he was motivated solely by India's independence. When that independence was achieved on August 15th, 1947, he was absent from the celebrations in New Delhi, as fame and glory did not drive him.

In the end, as often happens to extreme leaders, Gandhi's life was cut short by assassination. He was shot by a fanatic opposed to religious coexistence while his hands were folded in greeting at a prayer meeting. Remarkably, he blessed his assailant before collapsing, and he died on January 30th, 1948, at 5:12 PM.

**Key Observations of Gandhi's Philosophy:**

1. All people can shape and guide their lives according to the highest ideals, no matter how insignificant or powerless they may feel.

2. Be deeply rooted in your cultural and religious heritage while opposing all forms of social, ethnic, or religious intolerance.

3. Evil means will corrupt and degrade the purposes they are used for and the individuals who use them.

4. Personal change and the ability to bring about social change are linked to nonviolence, justice, celibacy (in thought and deed), and noncooperation.

Practice these principles in your own life first and then encourage society to adopt them. Individuals can create "zones of peace" in their lives by making every effort to banish violence, discord, and untruth.


**Additional Observations from Gandhi's Biography:**

- He meticulously accounted for every penny spent, recording every expenditure with a receipt.

- He believed that the heart's earnest and pure desires are always fulfilled.

- He saw Western civilization as predominantly based on force, unlike the Eastern.

- Despite living under British rule, he participated in the Boer War and served in an ambulance service in London during World War II, pursuing India's independence through the British Empire.

- He always contacted opposition leaders to allow them to respond before acting.

- He used wet earth bandages, or "good earth," to treat various ailments, including the black plague.

- He believed civility was the most challenging aspect of his philosophy, advocating for an inborn gentleness and desire to do good to opponents, not just the mere gentleness of speech cultivated for the occasion.

Gandhi had no idea how grand his vision would become when he began his journey. He believed it would achieve significant results and felt that, even in failure, it would cause no harm.

# RECAP

In this final chapter, we recap extreme leaders' crucial distinctions and duties, highlighting the bold vision, tough decisions, unwavering commitment, and essential skills needed to lead with extraordinary impact in extreme circumstances.

### Distinctions Between Leaders and Extreme Leaders

The distinctions between a leader and an extreme leader lie in the scope of their vision, the risks they are willing to undertake, and the moral foundation of their intent.

- **Scope of Vision**: An extreme leader possesses a grand vision transcending conventional leadership goals. While a typical leader aims for organizational success or incremental improvements, an extreme leader envisions transformative change that can impact entire communities, societies, or the world. Their vision often involves addressing deep-seated issues, challenging the status quo, and pioneering new paradigms.

- **Risks Involved**: Extreme leaders are characterized by their willingness to embrace significant risks. These risks are not just financial or strategic but often involve personal sacrifice, physical danger, and profound ethical dilemmas. Unlike conventional leaders who may play it safe to ensure steady progress, extreme leaders step into uncharted territories where failure could mean dire consequences. Their boldness and courage to face these risks head-on set them apart.

- **Moral Basis of Intent:** The moral foundation of an extreme leader's intent is paramount. A deep sense of justice, integrity, and ethical responsibility drives their actions. They are not merely focused on achieving success but are committed to doing so in a way that uplifts others, champions fairness, and adheres to high moral standards. This moral compass guides their decisions, even when facing tremendous pressure or temptation to compromise. However, history also shows that extreme leaders with immoral intentions can arise. Such leaders may possess a grand vision and take significant risks but are driven by harmful ideologies and unethical goals, leading to destructive outcomes. Despite evil intentions, their ability to galvanize followers through charisma and vision underscores the critical importance of moral integrity in leadership.

- **Impact on Followers**: Extreme leaders inspire unparalleled loyalty and dedication from their followers. The clarity of their vision and unwavering commitment to ethical principles galvanize people to join them in their quest. This deep connection often leads to extraordinary collective achievements.

- **Legacy and Influence**: The legacy of extreme leaders extends far beyond their immediate accomplishments. They leave an indelible mark on history, culture, and societal values. Their leadership often spawns movements, inspires future generations, and reshapes people's thinking about leadership and moral responsibility.

- **Adaptability and Resilience**: Extreme leaders demonstrate remarkable adaptability and resilience. They thrive in adversity, using setbacks as stepping-stones rather than stumbling blocks. Their ability to navigate complex, volatile environments with agility and poise is a testament to their exceptional leadership capabilities.

- **Ethical Leadership in Crisis**: In times of crisis, extreme leaders stand out by maintaining their ethical standards and moral clarity. While others may falter under pressure, these leaders uphold their principles, providing a steady hand and a guiding light through the turbulence. Their integrity during crises reinforces their followers' trust and solidifies their leadership.

The distinctions between a leader and an extreme leader are profound. Extreme leaders are defined by the expansive scope of their vision, their readiness to face extraordinary risks, and their moral intent—whether virtuous or malevolent. They are the architects of significant change, the beacons of leadership, and the embodiments of resilience and adaptability. By understanding and embracing these distinctions, we can better appreciate the unique qualities that make extreme leaders pivotal in shaping a better, more just world while remaining vigilant against those who might lead with harmful intentions.

### Checklist Extreme Leader Duties

The essence of extreme leadership involves several fundamental principles:

1. **Be the Leader**: Have a clear vision and set goals that support it.

2. **Involve the Team**: Engage followers in developing plans to achieve the goals.

3. **Lead by Example**: Demonstrate a strong work ethic and resilience through hardship.

4. **Delegate Wisely**: Assign the right tasks to the right team members.

5. **Respond Intelligently**: Address crises promptly and thoughtfully.

6. **Act Urgently**: Do not delay on urgent matters.

7. **Use Wisdom**: Make informed decisions based on experience.

8. **Remove Underminers**: Address dissent and remove those who threaten team cohesion.

9. **Implement Plans Quickly**: Get the team working on the plans they helped create as soon as possible.

10. **Monitor Progress**: Regularly check the team's progress and adapt as needed.

11. **Adapt and Be Flexible**: Adjust to changing circumstances and setbacks.

12. **Take Tough Action**: Confront and neutralize threats decisively.

13. **Prioritize the Team**: Take better care of the team than of oneself.

14. **Share the Glory**: Celebrate successes with the team.

15. **Accept Blame**: Take full responsibility for failures and setbacks.

16. **Admit Mistakes**: Be honest about errors and learn from them.

17. **Maintain Trust**: Always be truthful to protect the team's trust.

These principles and duties are not just guidelines, but the very building blocks that have the power to transform a good leader into an exceptional one, especially in extreme circumstances. They are the key to navigating the most challenging situations with grace and effectiveness.

### Strategies to Defeat an Extreme Leader

Defeating an extreme leader requires a multifaceted approach that leverages ethical scrutiny, undermines their vision, disrupts their risk-taking, erodes their follower base, challenges their adaptability, and capitalizes on crisis situations. By strategically addressing these aspects, it is possible to weaken the extreme leader's influence and ultimately defeat them. Here is what to do:

1. **Expose Ethical Flaws**:

    • *Integrity and Transparency*: Highlight any inconsistencies between the leader's actions and their proclaimed ethical standards. Investigate and publicize any unethical behavior or decisions.

    • *Media and Public Opinion*: Utilize media to expose these ethical flaws, swaying public opinion against the leader. Public scrutiny can erode their support base.

2. **Undermine Their Vision**:

    • *Counter-narratives*: Develop and promote a compelling counter-narrative that challenges the leader's vision. Offer a more practical and attainable vision that addresses the same issues without the associated risks.

• *Fact-Checking and Debunking*: Fact-check their claims and debunk myths or misinformation they spread. Provide clear, evidence-based alternatives to their proposed solutions.

3. **Disrupt Their Risk-Taking Capacity**:

• *Legal and Regulatory Pressure*: Use legal and regulatory mechanisms to limit their ability to take significant risks. This tactic can involve stricter enforcement of laws and regulations constraining their activities.

• *Highlight Consequences*: They should emphasize the potential negative consequences of their risk-taking, both to their followers and the broader public. This tactic can create doubt and reduce their followers' willingness to support risky ventures.

4. **Erode Their Follower Base**:

• Engage with Followers: Directly engage with their followers, addressing their concerns and offering alternative solutions. Build trust and demonstrate genuine commitment to their well-being.

• Promote Internal Dissent: Identify and support dissenting voices within the leader's group. Encourage critical thinking and debate, which can lead to fragmentation and weakening of the leader's support.

5. **Challenge Their Adaptability and Resilience**:

• *Create Unpredictable Challenges*: Introduce unexpected challenges and variables that require quick adaptation. This tactic can include political maneuvering, economic pressure, or social movements that disrupt their plans.

• *Sustain Pressure*: Apply sustained pressure that forces the leader to react and adapt constantly. This continuous strain can expose their limits and lead to mistakes.

6. **Leverage Crisis Situations**:

• *Ethical and Moral Emphasis*: In times of crisis, emphasize the need for ethical and moral leadership. Compare the leader's actions during the crisis with their ethical standards, highlighting any discrepancies.

• *Alternative Crisis Management*: Showcase alternative crisis management approaches that are more ethical, effective, and aligned with the public's values. This tactic can build trust in alternative leadership.

## Practical Implementation

1. **Building Alliances**: Form alliances with key stakeholders, including political entities, media organizations, and civil society groups. A unified front can amplify efforts to challenge the extreme leader.

2. **Strategic Communication**: Develop a strategic communication plan that includes social media campaigns, public statements, and grassroots engagement. Consistent messaging can influence public perception and weaken the leader's position.

3. **Resource Allocation**: Effectively allocate resources to support legal challenges, investigative journalism, and grassroots movements. Ensuring adequate funding and support can sustain long-term efforts against the extreme leader.

Effective implementation of these strategies requires a coordinated effort, leveraging the strengths of diverse stakeholders to undermine the extreme leader's influence systematically. By maintaining consistent communication and ensuring robust resource allocation, a unified front can achieve sustained pressure and ultimately lead to the leader's downfall.

-- The End --

# REFERENCES

Cohen, W. A. *The Art of the Leader*. Prentice Hall Trade, 1990.

"Followers Quotes." BrainyQuote, Xplore, www.brainyquote.com/quotes/keywords/followers.html. Accessed 15 June 2024.

Duggan, Bob, and Jeffery Yost. *Resilient Leadership*. Resilient Leadership LLC, 2010.

Fehrenbacher, D. E. *Selected Speeches and Writings/Lincoln*. Retrieved from http://showcase.netins.net/web/creative/lincoln/education/failures.htm.

Gandhi, M. K. *The Story of My Experiments with Truth*. Beacon Press, 1957.

Goleman, Daniel. *Emotional Intelligence: Why It Can Matter More Than IQ*. Bantam Books, 1995.

"How Did Mahatma Gandhi Spend August 15, 1947?" Asian Window, www.asianwindow.com/india/how-did-mahatma-gandhi-spend-august-15-1947/. Accessed 15 June 2024.

"How to Win Friends and Influence People." (1936). Retrieved from http://www.westegg.com/unmaintained/carnegie/win-friends.html.

Indian Child. "MAHATMA GANDHI Biography, information, Pictures." www.indianchild.com/mahatma_gandhi.htm. Accessed 15 June 2024.

Irving, W. *A History of the Life and Voyages of Christopher Columbus*. Jules Didot, Sr. and A. and W. Galignani, 1829.

Kahneman, Daniel. *Thinking, Fast and Slow*. Farrar, Straus, and Giroux, 2011.

Kane, D. K. *Arctic Explorations in the Years 1853, '54, '55*. Childs & Peterson, 1856.

Kotter, John P. *Leading Change*. Harvard Business School Press, 1996.

Northouse, Peter G. *Leadership: Theory and Practice*. 8th ed., Sage Publications, 2018.

"Northwest Passage." Retrieved from http://www.absolute-astronomy.com/topics/Northwest_Passage.

Patton, C. D. *Colt Terry, Green Beret*. Texas A&M University Press, 2005.

---. *Fifteen Secrets to Successful Timeshare Management*. Xlibris Corporation, 2009.

---. *Strategize Your Way to Success*. Xlibris Corporation, 2009.

---. *Strategies and Tactics*. Short Mystery Press, 2024.

Peale, N. V. *The Power of Positive Thinking*. Ballantine Books, 1996.

Pernoud, Regine, and Marie-Veronique Clin. *Joan of Arc: Her Story*. Translated by Jeremy Duquesnay Adams, St. Martin's Press, 1998.

Prescott, W. H. *History of the Conquest of Mexico*. John W. Lovell Company, 1843.

Twain, M. *Personal Recollections of Joan of Arc*. [Google Books]. Retrieved from books.google.com/books.

EXTREME LEADERSHIP

Wang, Dr. An. http://wang1200.org/history.html.

# ALSO BY CHARLES PATTON

**• For Honest Citizens Only**

A bold call for Americans to rise above politics and rebuild civic integrity.

**• In Defense of the Righteous**

A gripping story of moral courage when justice and survival collide.

**• Tigers of the Ice**

Adventure meets survival in an unforgiving world where instinct rules.

**• Mastering Strategy**

The essential guide to thinking, planning, and winning in any field.

**• Thinking**

Learn how to think more clearly, act decisively, and change your life.

**• Artificial Consciousness**

Explores the frontier of automating consciousness.

**• The Gardener's Secret and Other Stories**

Mysteries and dramas revealing the hidden motives behind ordinary lives.

**• Who Do You Trust**

A deadly game of deceit between two spies and one truth.

**• Busted, What's Wrong With My Excuse**

An entertaining look at excuses people make, and how to excuse better.

**• Naked Reflections**

Raw, honest poetry of truth, ego, and the search for authenticity.

**• Charles Patton, Visionaire**

Insights from a lifetime of ideas, invention, and fearless creativity.

**• Storming the Castle Bridge**

A tale of rebellion, loyalty, and the unbreakable human will to be free.

*Find every title at: charlespattonbooks.com*